Lecture Notes in Mathematics

Edited by A. Dold and B. Eckmann

651

T0239446

Peter W. Michor

Functors and Categories of Banach Spaces

Tensor Products, Operator Ideals and Functors on Categories of Banach Spaces

Springer-Verlag
Berlin Heidelberg New York 1978

Author
Peter W. Michor
Institut für Mathematik
Universität Wien
Strudlhofgasse 4
A-1090 Wien

Library of Congress Cataloging in Publication Data

Michor, Peter W 1949-
Functors and categories of Banach spaces.

(Lecture notes in mathematics ; 651)
Bibliography: p.
Includes index.
1. Banach spaces. 2. Categories (Mathematics)
3. Functor theory. I. Title. II. Series: Lecture notes in mathematics (Berlin) ; 651.
QA3.L28 no. 651 [QA322.2] 510'.8s [515'.73] 78-6814
ISBN 0-387-08764-8

AMS Subject Classifications (1970): 46M05, 46M15

ISBN 3-540-08764-8 Springer-Verlag Berlin Heidelberg New York
ISBN 0-387-08764-8 Springer-Verlag New York Heidelberg Berlin

Printing and binding: Beltz Offsetdruck, Hemsbach/Bergstr.
2141/3140-543210

Preface *)

The aim of this book is to develop the theory of Banach operator ideals and metric tensor products along categorical lines: these two classes of mathematical objects are endofunctors on the category Ban of all Banach spaces in a natural way and may easily be characterized among them (§4). Up to now they were investigated with methods of functional analysis in a sort of ad hoc manner and with an outlook to special properties; here they are subject to several categorical and universal constructions:

Kan extensions from the subcategory of finite dimensional spaces are studied in §2 and applied to tensor products and operator ideals in §§ 4,5,6 and give rise to the reappearance of the ⊗-norms in the sense of Grothendieck and to minimal and maximal operator ideals in the sense of Pietsch.

Duality for co- and contravariant functors is studied in §3 (and some new and deep results are derived on it) and is applied to tensor products and operator ideals in §§ 4,5,6: duality is the link between the two notions.

Several other constructions of sub- and quotient functors induced by canonical adjoint relations are used to (co) reflect all appearing functors back to tensor products and operator ideals (§§4,5,6).

In §7 we introduce (as an example) a new class of tensor products, the projective (p,r,s)-tensor product, which is a link between the (p,r,s)-absolutely summing, - nuclear and - integral operator ideals and we use it to derive a lot of new relations between these operator ideals from existing ones.

The whole subject - although sometimes technical and complicated - seems to be a succussful and deep application of category theory to functional analysis.

P.M.

*) Research was partially done while visiting the University of Warwick supported by a Royal Society award.

TABLE OF CONTENTS

PAGE

§ 0 INTRODUCTION

Operator ideals in Hilbert spaces caused mathematical interest since the first days of functional analysis; attention was focused on the behaviour of tensor products and operator ideals on Banach spaces thirty years ago and the intimate connection between them is clear since then. Operator ideals showed to be more tractable and more work was invested recently to research them.

Tensor products and operator ideals are in fact special examples of bifunctors on the category of Banach spaces and it is maybe worthwhile to look what theoretical setting the category theory provides for them: they can easily be characterized among other bifunctors by certain functorial properties using some elementary adjunctions (§ 4), and the characterization seems to me to be simpler than the original definition.

The connection between tensor products and operator ideals turns out to coincide with the notion of duality of functors, the Banach space analog for the Eckmann-Hilton - duality, and it appeared in MITJAGIN-SHVARTS [16] some ten years ago. To apply it we show some deep results on the duality of functors, in fact, we compute the dual of any functor of type Σ , and we give a short account of an appropriate duality theory for contravariant functors too (§ 3).

The functorial analog for the notion of tensor norm of GROTHENDIECK [8] is explained in § 2. Its importance for the duality of functors was first pointed out by HERZ-PELLETIER [9], who called it computability. We give a new approach to computable functors and derive some new results on them using heavily the tensor product of functors and the "exponential law" that goes along with it, due to CIGLER [3]. This gives us at hand a formal machinery for a nearly purely algebraic handling with functors, tensor products and spaces of natural transformations.

Computable bifunctors of type Σ correspond exactly to the tensor norms of GROTHENDIECK [8], and an operator ideal is minimal in the sense of PIETSCH [19] if it is computable whenever the Banach spaces considered have the metric approximation property; in the general case a slight factorizationg links the two notions (§ 5).

The notion dual to computability is that of complete functors, which appeared first in CIGLER [4] in a special case. The completion of the identity functors is the bidual functor '' ; computable bifunctors and maximal operator ideals differ inasmuch as in the former always appears a bidual space: repairing that by taking a pushout links the two notions (§ 6).

In § 7 we introduce a new class of tensor norms, the projective (p.r.s) - tensor product, which generalizes the p-tensor products of SAPHAR [20] and CHEVET [2]. Its dual functor is the ideal of (p.r.s) - absolutely summing operators, which we define a little different from PIETSCH [19]; its associated operator ideal is that of (p.r.s) - nuclear operators, its dual tensor norm give the inductive (p.r.s)- tensor product, whose dual functor again is the ideal of (p.r.s) integral operators. By this theoretical interdependence we are able to carry over to all these functors results hitherto known only for (p.r.s) - absolutely summing operators.

We limited ourselves to normed operator ideals, and we take always convex hulls of unit balls whenever quasinorms appear in § 7, but surely we lost information by that process. To produce the right background for the theory of quasinormed operator ideals as developed in PIETSCH [17], we should provide a theory of functors from the category of Banach spaces into the category of complete quasinormed spaces; this would amount to a study of quasi-tensor-norms.

The second limitation of this work is that we did not embody the notion of injective or projective tensor-norms or operator ideals as they appear in GROTHENDIECK [8] and PIETSCH [17]. The reason is that we were not able to find an analog of theorem 3.4 of SAPHAR [20], which would connect (p.r.s) - integral and (p.r.s) - factorizable operators.

Besides the facts of the theory of categories of Banach spaces, listed in § 1, and some moderate abstract category theory (chapters I and IV of MACLANE [12] suffice) we presuppose only knowledge in functional analysis.

§1. PRELIMINARIES

Let Ban be the category whose objects are a big enough class of Banach spaces and whose morphisms are bounded linear maps; Ban_1 we define to consist of the same class of objects, but we admit only contractive linear maps as morphisms, i.e. linear maps with norm ≤ 1. Ban is an addive Ban-based category in terms of relative category theory, but Ban_1 has the advantage that it is complete and co-complete, i.e. contains all limits and co-limits of small spectral families. Thus most of the time we will regard Ban_1 and Ban together, but we understand that only contractive morphisms are relevant if we speak of limits, colimists and other universal concepts. By using a equality sign we understand always that this is an isometric isomorphism, we will always very strictly distinguish between isomorphisms and isometric isomorphisms. The ground field I is R or C, we do not specify, but only of them.

1.1. By $H(X,Y)$ we design the Banach space of all bounded linear maps $X \to Y$. By a (covariant) functor F: Ban $\to$ Ban we mean a map that associates new Banach spaces $F(X)$ to old ones X and associates a morphism $F(f):F(X) \to F(Y)$ to each morphism $f:X \to Y$ in such a way that the transformation $f \to F(f)$ from $H(X,Y)$ into $H(F(X),F(Y))$ is contractive and linear and the usual functional properties hold:

$F(f\circ g) = F(f) \circ F(g)$ and $F(1_X) = 1_{F(X)}$. A contravariant functor $\overline{F}$: Ban → Ban then transforms $f: X \to Y$ into $\overline{F}(f): \overline{F}(Y) \to \overline{F}(X)$ and $\overline{F}(f\circ g) = \overline{F}(g) \circ \overline{F}(f)$; all other properties are the same as those of a covariant functor.

1.2. The simplest examples of functors are the following: $H(X,.)$, $H(X,f)(g) = f\circ g$, is the covariant partial functor of the Hom functor of Ban. $H(.,X)$, $H(f,X)(g) = g\circ f$, is the contravariant part of it.

For $X,Y \in$ Ban let $X \hat{\otimes} Y$ be the completion of the algebraic tensor product $X \otimes Y$ in the greatest crossnorm.

$$\| u \|^{\wedge} = \inf \sum_{i=1}^{n} \| x_i \| \; \| y_i \| \text{, where } u = \sum_{i=1}^{n} x_i \otimes y_i \text{ runs}$$

through all representations of u in $X \otimes Y$. $\hat{\otimes}$. is a co-covariant bifunctor, its action on morphisms is given by

$$(f \hat{\otimes} g) \; (\Sigma \; x_i \otimes y_i) = \Sigma \; f(x_i) \otimes g(y_i).$$

By $X \check{\otimes} Y$ let us denote the closure of the algebraic tensor product $X \otimes Y$ in $H(X',Y)$ via the embedding $X \otimes Y \to H(X',Y)$, given by $\Sigma \; x_i \otimes y_i \to (x' \to \Sigma \langle x_i, x'\rangle y_i)$.

Its norm is given by

$$\begin{aligned} \| \Sigma \; x_i \otimes y_i \|^{\vee} &= \sup_{\|x'\| \leq 1, \|y'\| \leq 1} |\Sigma \langle x_i, x'\rangle \langle y_i, y'\rangle| \\ &= \sup_{\|x'\| \leq 1} \| \Sigma \langle x_i, x'\rangle y_i \|_Y \\ &= \sup_{\|y'\| \leq 1} \| \Sigma \; x_i \langle y_i, y'\rangle \|_X . \end{aligned}$$

Among the first who studied functors on the category of Banach spaces were MITJAGIN-SVARTS [16]; With respect to tensor products of Banach spaces we refer the reader to SCHATTEN [21] and GROTHENDIECK [8].

1.3. A natural transformation η from the functor F into another one F_1 is a family of morphisms $(\eta_X)_{X \in \mathrm{Ban.}}$ where $\eta_X \in H(F(X), F_1(X))$ such that for any $f \in H(X,Y)$ the diagram

$$\begin{array}{ccc} F(X) & \xrightarrow{\ \eta_X\ } & F_1(X) \\ {\scriptstyle F(f)}\big\downarrow & & \big\downarrow{\scriptstyle F_1(f)} \\ F(Y) & \xrightarrow{\ \eta_Y\ } & F_1(Y) \end{array}$$

commutes

and furthermore $\|\eta\| = \sup_X \|\eta_X\| < \infty$ holds.

The class of all natural transformations $F \to F_1$ is a Banach space which we denote by Nat (F,F_1) if it is a set. In most cases it is a set and we pay no attention whether this is so in general. See the general investigation of this (LINTON [11] etc.).

1.4. The projective tensor product $X \hat{\otimes} Y$ of X and Y has the following universal property: given any bounded bilinear map $\varphi : X \times Y \to Z$ into an arbitrary Banach space Z then there is a unique linear map

$\hat{\varphi} : X \hat{\otimes} Y \to Z$ with

$\|\hat{\varphi}\| \leq \|\varphi\|$ and $\varphi = \hat{\varphi} \circ \Pi$

where $\Pi : X \times Y \to X \hat{\otimes} Y$ is

$$\begin{array}{ccc} X \times Y & \xrightarrow{\ \Pi\ } & X \hat{\otimes} Y \\ {\scriptstyle \varphi}\big\downarrow & \swarrow {\scriptstyle \hat{\varphi}} & \\ Z & & \end{array}$$

the canonical bilinear map $\Pi(x,y) = x\otimes y$.

Using this properly we see very easily that

$H(X \hat{\otimes} Y, Z) = H(X,H(Y,Z))$ holds natural in X,Y,Z, i.e. the equality sign is an invertibleisometrical natural transformation of trifunctors

$$H(.\hat{\otimes}..,\ ...) = H(.,H(..,...)).$$

Another way to express this fact is to say that the functor $Y\hat{\otimes}.$ is left adjoint to $H(Y, .)$ and that the adjunction is natural in Y.

By a general category theoretical result $Y\hat{\otimes}.$ comutes thus with colimits in Banach and $H(Y, .)$ with limits; special cases are

$$Y\hat{\otimes}\ell^1_s = \ell^1_s(Y) \quad \text{and} \quad \ell^\infty_s(Y') = \ell^\infty_s(H(Y,I)) =$$

$$= H(Y,\ell^\infty_s) = H(Y,\ell^1_s{}') = H(Y,H(\ell^1_s,I)) =$$

$$= H(\ell^1_s\hat{\otimes} Y, I) = (\ell^1_s(Y))'.$$

1.5. On the other hand the projective tensor product $.\hat{\otimes}.$ is uniquely determined by its property to comute with colimits in Ban_1 and hence by its property to be a left adjoint: Every Banach space X may be canonically represented as a colimit in Ban_1 of a spectral family consisting of finite dimensional spaces of the form ℓ^1_n, where n stands for $\{1,...,n\}$ (see CIGLER [5], Page 15). Now let F be a functor which comutes with colimits, let be

$X \in$ Ban, $X = \varinjlim \{\ell_n^1\}$. Then we have

$$F(X) = F(\varinjlim \{\ell_n^1\}) = \varinjlim \{F(\ell_n^1)\} = \varinjlim \{\ell_n^1(F(I))\} =$$
$$= \varinjlim \{\ell_n^1 \hat{\otimes} F(I)\} = (\varinjlim \{\ell_n^1\}) \hat{\otimes} F(I) = X \hat{\otimes} F(I).$$

That is the essential content of the paper of SEMADENI-WIWEGER [22].

1.6. Now let us consider a contravariant functor $\overline{F}$: $\mathrm{Ban}^{op} \to$ Ban and a covariant one F: Ban → Ban. We restrict them to some subcategory $\underline{C}$ of Ban and define the tensorproduct of $\overline{F}$ and F over $\underline{C}$ in the following way:

A dinatural transformation α of the bifunctor $\overline{F}(.) \hat{\otimes} F(..)$ into a Banach space Z is a family $(\alpha_X)_{X \in \underline{C}}$ of morphisms α_X: $\overline{F}(X) \hat{\otimes} F(X) \to Z$ such that for each $f \in H(X,Y)$ the following diagram comutes and moreover $\|\alpha\| = \sup_X \|\alpha_X\| < \infty$ holds:

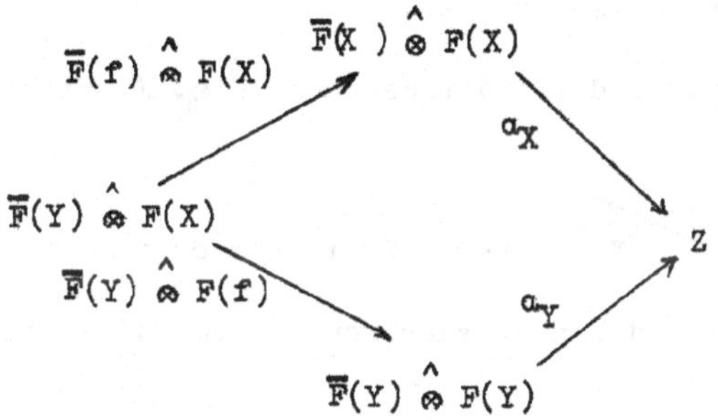

It is easy to see that a family (α_X) defines a dinatural transformation if and only if it corresponds to a family (β_X) of morphisms $F(X) \to H(\overline{F}(X),Z)$ under the isomorphism

$H(\overline{F}(X) \hat{\otimes} F(X),Z) = H(F(X), H(\overline{F}(X),Z))$, which defines a natural transformation $F \to H(\overline{F}(.), Z)$.

Now the $\underline{C}$- tensorproduct of $\overline{F}$ and F is a Banach space which we denote by $\overline{F} \hat{\otimes}_{\underline{C}} F$ together with a dinatural map

$\Pi : \overline{F}(.) \hat{\otimes} F(..) \to \overline{F} \hat{\otimes}_{\underline{C}} F$

such that each dinatural map $\varphi : \overline{F}(.) \hat{\otimes} F(..) \to Z$ into an arbitrary Banach space Z

$$\begin{array}{ccc} \overline{F}(.) \hat{\otimes} F(..) & \xrightarrow{\ \Pi\ } & \overline{F} \hat{\otimes}_{\underline{C}} F \\ \downarrow \varphi & \swarrow \hat{\varphi} & \\ Z & & \end{array}$$

factors uniquely over Π to $\hat{\varphi} : \overline{F} \hat{\otimes}_{\underline{C}} F \to Z$ with $\|\hat{\varphi}\| \leq \|\varphi\|$.

By universal reason ing (some people might say: by abstract nonsense) the $\underline{C}$ - tensorproduct of $\overline{F}$ and F is uniquely determined up to an isometric isomorphism.

If we disregard set-theoretical difficulties then $\overline{F} \hat{\otimes}_{\underline{C}} F$ exists and is given by

$$\overline{F} \hat{\otimes}_{C} F = \sum_{X \in C} \overline{F}(X) \hat{\otimes} F(X) \Big/ N$$

, where Σ is the coproduct in Banach and N is the closed linear subspace generated by all elements of the form

$$\sum_K \overline{F}(\varphi) y_K \otimes x_K - \sum_K y_K \otimes F(\varphi)\, x_K ,$$

where $\sum_K y_K \otimes x_K \in \overline{F}(Y) \hat{\otimes} F(X)$ and $\varphi \in H(X,Y)$.

The notion of the $\underline{C}$ - tensor product and this representation here is due to CIGLER [3].

1.7. The $\underline{C}$ - tensor product $\overline{F} \hat{\otimes}_{\underline{C}} F$ is the colimit in Ban_1 of the following spectral family: let $\underline{C}_1$ be $\underline{C} \cap Ban_1$. Then the index category of the spectral family is the so called twisted morphism category of $\underline{C}_1$, i.e. indices of the spectral family are all isomorphisms in $\underline{C}_1$, to each $f \in \underline{C}_1(X,Y)$ we assign the space $R^f = \overline{F}(Y) \hat{\otimes} F(X)$ and to each commutative diagram

$$\begin{array}{ccc} X & \xrightarrow{f} & Y \\ g \downarrow & & \uparrow h \\ X_1 & \xrightarrow{f_1} & Y_1 \end{array}$$

in $\underline{C}_1$ we assign a morphism $\pi(g,h) = \overline{F}(h) \hat{\otimes} F(g) : R^f \to R^{f_1}$.

Then $\overline{F} \hat{\otimes}_{\underline{C}} F = \underline{\lim} \{R^f\}$, see MICHOR [13].

1.8. Since dinatural transformations $\overline{F}(.) \hat{\otimes} F(..) \to Z$ and natural transformations $F \to H(\overline{F}(.), Z)$ correspond to each other uniquely and isometrically, we see immediately that

$H(\overline{F} \hat{\otimes}_{\underline{C}} F, Z) = Nat(F, H(\overline{F}(.), Z))$ holds. Since moreover $\overline{F} \hat{\otimes}_{\underline{C}} F$ is a natural construction, i.e. natural transformations $\varphi : \overline{F} \to \overline{F}_1$ and $\psi : F \to F_1$ induce a map

$\psi \hat{\otimes}_{\underline{C}} \psi : \overline{F} \hat{\otimes}_{\underline{C}} F \to \overline{F}_1 \hat{\otimes}_{\underline{C}} F_1$

it is straightforward to check that the following general "exponential law" holds:

Let $\underline{C}$ and $\underline{D}$ be subcategories of Ban, let $M : \underline{C}^{op} \times \underline{D} \to Ban$ be a contra-covariant bifunctor and $F : \underline{C} \to Ban$, $F_1 : \underline{D} \to Ban$ be covariant functors. Then

$$\underset{(..)\in \underline{D}}{\mathrm{Nat}} \left(M(.,..) \underset{(.)\in \underline{C}}{\hat{\otimes}} F(.) \, , F_1(..)\right) =$$

$$= \underset{(.)\in \underline{C}}{\mathrm{Nat}} \left(F(.) \, , \underset{(..)\in \underline{D}}{\mathrm{Nat}} (M(.,..) \, , F_1(..))\right)$$

holds naturally in F , M and F_1 .

See CIGLER [5] for a detailed discussion.

A special case is the following:

$$(\overline{F} \hat{\otimes}_{\underline{C}} F)' = H(\overline{F} \hat{\otimes}_{\underline{C}} F \, , I) = \underset{\underline{C}}{\mathrm{Nat}} (F \, , (\overline{F})') \, .$$

1.9. Let A be a Banach space. Then we have

$$\mathrm{Nat}\,(A \hat{\otimes} . , F) = H(A \, , F(I)) \, , \text{ where}$$

F : Ban $\to$ Ban is any functor, is given by

$\varphi \longrightarrow \varphi_I$, $\varphi_X (a \otimes x) = F(\hat{x}) \circ \varphi_I$ where

$\hat{x} \in H(I \, , X)$ corresponds to $x \in X$ by $\hat{x}(r) = r \cdot x$.

Thus the functor that assigns $A \hat{\otimes} .$ to A is left adjoint to the forgetful functor $F \longmapsto F(I)$.

The counit ε of this adjunction is given by

$$\varepsilon_X^F : F(I) \hat{\otimes} X \longrightarrow F(X) \, ,$$

$$\varepsilon_X^F (\Sigma\, a_i \otimes x_i) = \Sigma\, F(\hat{x}_i)\, a_i \, .$$

Whenever the image of ε_X^F is dense in $F(X)$ for all X we say that the functor F is of type Σ or essential. LEVIN [10] proved that for all functors F the restriction of ε_X^F to the algebraic tensorproduct $F(I) \otimes X$ is injective and that $\|.\|_{F(X)}$ induces on the subspace $X \otimes F(I)$ a reasonable crossnorm in the sense of GROTHENDIECK [8].

Since ε_X^F is natural in X the closure of the image of ε_X^F, i.e. the closure of $F(I)\hat{\otimes} X$ in $F(X)$, defines a subfactor of F, which we denote by F_e and we call it the essential part or the partial functor of type Σ of F.

$F_e(X) = X\otimes_\alpha F(I)$ is therefore the completion of $X\otimes_\alpha F(I)$ in a reasonable crossnorm, i.e. a norm α on $X\otimes F(I)$ which satisfies $\|.\|^\wedge \geq \alpha \geq \|.\|^{\hat{\wedge}}$, and this norm α is functorial in X: given $f : X \to Y$ then $f \otimes_\alpha F(I)$: $X\otimes_\alpha F(I) \longrightarrow Y\otimes_\alpha F(I)$ is a map with norm $\leq \|f\|$.

1.10. The analogous notion exists for contravariant functors $\bar{F}$: Ban $\to$ Ban.

Nat $(A \hat{\otimes} .' , \bar{F}) = H(A,\bar{F}(I))$ holds naturally in A and $\bar{F}$, the counit of this adjunction is given by

$$\varepsilon_X^{\bar{F}} : \bar{F}(I) \hat{\otimes} X' \to \bar{F}(X)$$

$$\varepsilon_X^{\bar{F}} (\Sigma a_i \otimes x_i') = \Sigma \bar{F}(x_i') a_i .$$

Again $\varepsilon_X^{\bar{F}} \mid \bar{F}(I) \otimes X'$ is injective and $\|.\|_{\bar{F}(X)}$ induces a reasonable crossnorm α on $\bar{F}(I) \otimes X'$ which is functorial in X; i.e. given $f : X \to Y$, then $\bar{F}(I) \otimes f' : \bar{F}(I) \otimes_\alpha Y' \to \bar{F}(I) \otimes_\alpha X'$ has norm $\leq \|f\|$. Whether α is functorial in X' too, i.e. given any $g : X' \to Y'$ (even non weak-$*$ continuous ones) the question whether

$\overline{F}(I) \hat{\otimes}_\alpha g : \overline{F}(I) \hat{\otimes}_\alpha X' \to \overline{F}(I) \hat{\otimes}_\alpha Y'$ has norm $\leq \|g\|$ too will be one of the topics of this article. Again the closure of $\overline{F}(I) \otimes X'$ in $\overline{F}(X)$ defines a contravariant subfunctor of $\overline{F}$, which we again design by $\overline{F}_e$ and we call it again the essential part or the subfunctor of type Σ.

1.11. Any natural transformation maps essential parts of functors into essential parts, since the counit ε^F is natural in F too, i.e. $\mathrm{Nat}\,(F,F_1) = \mathrm{Nat}\,(F,F_{1e})$ if F is of type Σ.

1.12. A Banach space X is said to have the (metric) approximation property, if for each compact subset $K \subseteq X$ and $\varepsilon > 0$ there is a bounded linear map $u : X \to X$ of finite rank (with $\|u\| \leq 1$) such that $\|u(x) - x\| \leq \varepsilon$ holds for all $x \in K$. It is easy to see that X has the (metric) approximation property, if and only if the Banach algebra $K(X,X)$ of all compact linear maps $X \to X$ has a left approximate identity (bounded by one) consisting of maps of finite rank.

Consider the following diagram

$$\begin{array}{ccc} X \hat{\otimes} Y & \xrightarrow{\quad s \quad} & L(X',Y) \\ {\scriptstyle \text{coims}} \downarrow & & \uparrow {\scriptstyle \text{ims}} \\ L^1(X',Y) & \xrightarrow{\quad \tilde{s} \quad} & X \hat{\hat{\otimes}} Y \end{array},$$

where $L(X',Y)$ is the space of all linear maps whose restriction to the unit ball OX' of X' is weak* $- \|\cdot\|$ - continuous, s is the map $x \otimes y \to (x' \to \langle x,x'\rangle y)$, $L^1(X',Y) = X \hat{\otimes} Y / s^{-1}(o)$ and

$X \hat{\hat{\otimes}} Y$ is the closure of the image of s.

X has the approximation property if and only if coims is injective for all Banach spaces Y (i.e. $X \hat{\otimes} Y = L^1(X',Y)$), if and only if ims is surjective for all Banach spaces Y (i.e. $X \hat{\hat{\otimes}} Y = L(X',Y)$).

For further information see GROTHENDIECK [7]. The first example of a Banach space without the approximation property is due to ENFLO [23].

<u>1.13.</u> Special results and examples of tensor products of functors.

(a) $H(.,X) \hat{\otimes}_{\underline{C}} F = F(X)$ naturally in F in X whenever $\underline{C}$ in a full subcategory of Ban that contains X. This corresponds to the Yonedalemma

Nat $(H(X,.),F) = F(X)$. See CIGLER [3].

(b) $\overline{F} \hat{\otimes}_{\underline{C}} H(X,.) = \overline{F}(X)$ holds under the same restrictions.

(c) $(.' \hat{\hat{\otimes}} X) \hat{\otimes}_{Ban} F = F_e(X)$ holds whenever X has the metric approximation property. The proof relies on the existence of a bounded left approximate identity in $K(X,X) = X' \hat{\hat{\otimes}} X$. See CIGLER [3].

(d) $\overline{F} \hat{\otimes}_{Ban} (X' \hat{\hat{\otimes}} .) = \overline{F}_e(X)$ holds whenever X' has the metric approximation property. Here we would require a right approximate identity, so the proof is more complicated. See MICHOR [15].

(e) $\overline{F} \hat{\otimes}_{Ban} F = \overline{F}(I) \otimes_{\alpha} F(I)$ for a reasonable crossnorm α whenever $\overline{F}$ or F is of type Σ. See MICHOR [13].

(f) $$(\overline{F}(.) \hat{\otimes}_{(.)\in \underline{C}} M(..,.)) \hat{\otimes}_{(..)\in \underline{D}} F(..) =$$

$$= \overline{F}(.) \hat{\otimes}_{(.)\in \underline{C}} (M(..,.) \hat{\otimes}_{(..)\in \underline{D}} F(..))$$

holds for any contra-covariant bifunctor $M : \underline{D}^{op} \times \underline{C} \to$ Ban, as can be proved by showing that the adjoint of the obvious map is isometric onto using the exponential law. See CIGLER [5].

(g) $(X \hat{\otimes} .') \hat{\otimes}_{Ban} F = X \hat{\otimes} F(I)$ and $\overline{F} \hat{\otimes}_{Ban} (X \hat{\otimes} .) = X \hat{\otimes} \overline{F}(I)$ hold, see MICHOR [13].

§2. Computable functors

2.1. Remark: Besides the notion of reasonable crossnorm, that is essentially due to SCHATTEN [21], GROTHENDIECK [8] introduced the concept of a so called $\otimes$-norm, i.e. a bifunctorial crossnorm α that satisfies the following condition: for $u \in X \otimes Y$

$\alpha(u) = \inf ||u||_{E \otimes_\alpha F}$ holds, where E,F run through all finite dimensional subspaces of X,Y respectively. HERZ-PELLETIER [9] saw that this notion is useful for computing the dual functor of a functor (see next section) and called it computability.

2.2. Let Fin be the full subcategory of all finite dimensional Banach spaces in Ban. Given a functor F: Fin $\to$ Ban and $X \in$ Ban, represent X as colimit in Ban_1 of all its finite dimensional subspaces, $X = \lim\limits_{\to} \{E, E \subset X, E \in Fin\}$, and consider the Banach space

$$LF(X) = \lim_{\to} \{F(E), E \subset X, E \in Fin\}.$$

By the universal property of colimits it is very readily seen that $X \mapsto LF(X)$ is the object transformation of a functor LF: Ban $\to$ Ban, that L: $Ban^{Fin} \to Ban^{Ban}$ is a functor and that L is left adjoint to the restriction functor $F \to F/Fin$ from Ban^{Ban} into Ban^{Fin}, i.e.

$$\underset{Fin}{Nat}\,(F, F_1|Fin) = \underset{Ban}{Nat}\,(LF, F_1) \text{ holds naturally in F and } F_1 .$$

See HERZ-PELLETIER [9] for that.

Clearly $(LF)|Fin = F$ holds for any F:Fin $\to$ Ban, but the counit $L(F_1|Fin) \to F_1$ of this adjunction is no equivalence.

We say that F: Ban $\to$ Ban is a computable functor if $L(F|Fin) = F$ holds.

2.3. Proposition: HERZ-PELLETIER [9]

If F: Ban → Ban is computable then

$$\operatorname*{Nat}_{Ban}(F,F_1) = \operatorname*{Nat}_{Fin}(F|Fin, F_1|Fin) \text{ for all functors } F_1 : Ban \to Ban.$$

Proof:
$$\operatorname*{Nat}_{Ban}(F,F_1) = \operatorname*{Nat}_{Ban}(L(F|Fin),F_1)$$

$$= \operatorname*{Nat}_{Fin}(F|Fin,F_1).$$

2.4. Proposition: For any F: Fin → Ban we have

$$LF(.) = H(..,.)\ \hat{\otimes}_{(..)\in Fin}\ F(..) = (..' \hat{\hat{\otimes}} .)\ \hat{\otimes}_{(..)\in Fin}\ F(..).$$

Thus: F:Ban → Ban is computable if and only if $F(X) = (.' \hat{\hat{\otimes}} X)\ \hat{\otimes}_{Fin}\ F$ holds for all $X \in$ Ban.

Proof: Define $L'F$ by $L'F(X) = (.' \hat{\hat{\otimes}} X)\ \hat{\otimes}_{Fin}\ F$, then $L'F$ is clearly a functor and L' is that too by the discussion in 1.8. Using the exponential law 1.8 we see that L' is left adjoint to the restriction functor:

$$\operatorname*{Nat}_{(.)\in Ban}\left(H(..,.)\ \hat{\otimes}_{(..)\in Fin}\ F(..),\ F_1(.)\right)$$

$$= \operatorname*{Nat}_{(..)\in Fin}\left(F(..),\ \operatorname*{Nat}_{(.)\in Ban}(H(..,.),F_1(.))\right)$$

$$= \operatorname*{Nat}_{(..)\in Fin}(F(..),F_1(..)) \text{ by the Yoneda lemma}$$

$$= \operatorname*{Nat}_{Fin}(F,F_1|Fin).$$

The naturality of this relation follows from the naturality of the exponential law.

Since any two left adjoints of the same functor are naturally equivalent, we see that $L = L'$ holds naturally and isometrically.

2.5. Proposition: For any functor F: Ban → Ban we have

$$LF(X) = (.'\hat{\otimes}X)\underset{\text{Fin}}{\hat{\otimes}} F = (.'\hat{\otimes}X)\underset{\text{Ban}}{\hat{\otimes}} F.$$

Proof: By a little abuse of notation we wrote LF for $L(F|\text{Fin})$. Clearly we have a canonical map $(.'\hat{\otimes}X)\underset{\text{Fin}}{\hat{\otimes}} F \to (.'\hat{\otimes}X)\underset{\text{Ban}}{\hat{\otimes}} F$ (restrict the canonical dinatural map $\Pi: (.'\hat{\otimes}X)\hat{\otimes} F(..) \to (.'\hat{\otimes}X)\underset{\text{Ban}}{\hat{\otimes}} F$. to the subcategory Fin and use the universal property 1.6). Its adjoint is easily seen to be the following isometrical isomorphism, thus this map is one too.

$$((.'\hat{\otimes}X)\underset{\text{Ban}}{\hat{\otimes}} F)' = \underset{\text{Ban}}{\text{Nat}}\,(.'\hat{\otimes}X, F(.)') \text{ by 1.8.}$$

$$= \underset{\text{Fin}}{\text{Nat}}\,(.'\hat{\otimes}X, F(.)') \text{ by 2.10 and 2.13(b) below}$$

$$= ((.'\hat{\otimes}X)\underset{\text{Fin}}{\hat{\otimes}} F)'. \quad \text{qed.}$$

2.6. Corollary: LF is of type Σ for any functor F: Ban → Ban.

Proof: $X \otimes F(I)$ is dense in $(.'\hat{\otimes}X)\underset{\text{Ban}}{\hat{\otimes}} F$ by 1.13(e).

2.7. Examples:

(a) Clearly $X\hat{\otimes}.$ is computable since it comutes with any colimits.

(b) $X\hat{\hat{\otimes}}.$ is computable too, as can be checked by routine computation (HERZ-PELLETIER [9]).

(c) Thus LF is computable for any functor F:Ban → Ban, since we can proceed as follows:

$$L(LF)(X) = (.' \hat{\hat{\otimes}} X) \underset{(.)\in Ban}{\hat{\otimes}} ((..' \hat{\hat{\otimes}}.) \underset{(..)\in Ban}{\hat{\otimes}} F(..))$$

$$= ((.' \hat{\hat{\otimes}} X) \underset{(.)\in Ban}{\hat{\otimes}} (..' \hat{\hat{\otimes}}.)) \underset{(..)\in Ban}{\hat{\otimes}} F \text{, by 1.11(g),}$$

$$= (..' \hat{\hat{\otimes}} X) \underset{(..)\in Ban}{\hat{\otimes}} F \text{, since } Y' \hat{\hat{\otimes}}. \text{ is computable,}$$

$$= LF(X).$$

(d) If X is a Banachspace without the approximation property, then the functor $L^1(X',.)$ of 1.11 is not computable, since it agrees with computable functor $X\hat{\otimes}.$ on Fin and thus

$L(L^1(X',.)) = X\hat{\otimes}. \neq L^1(X',.)$ on Ban.

This last counterexample is typical, as the following proposition shows.

<u>2.8.</u> <u>Proposition</u>: <u>If</u> F: Ban → Ban <u>is any functor and</u> X <u>has the metric approximation property, then</u> $LF(X) = F_e(X)$.

Thus the computable functors $\underline{A} \to$ Ban are exactly those of type Σ, where $\underline{A}$ is the full subcategory of Ban consisting of all Banachspaces with the metric approximation property.

<u>Proof</u>: $LF(X) = (.' \hat{\hat{\otimes}} X) \underset{Ban}{\hat{\otimes}} F$ by 2.4

$= F_e(X)$ by 1.13c).

<u>2.9.</u> Since we will deal later on with operator ideals and these have contravariant partial functors we will need a notion of computability for contravariant functors too.

We consider now the category $Ban^{Ban^{op}}$ of contravariant functors $\bar{F}$: Ban → Ban and the restriction functor $\bar{F} \to \bar{F}|Fin$ from the category $Ban^{Ban^{op}}$ into $Ban^{Fin^{op}}$.

Proposition: The restriction functor $\bar{F} \to \bar{F}|Fin$ for contravariant functors $\bar{F}: Ban^{op} \to Ban$ has a left adjoint $L: Ban^{Fin^{op}} \to Ban^{Ban^{op}}$. For $\bar{F}: Fin^{op} \to Ban$ $L\bar{F}$ is given by $L\bar{F}(X) = \bar{F}(.) \hat{\otimes}_{Fin} H(X,.)$.

Proof: $L\bar{F}: Ban^{op} \to Ban$ is clearly a functor and $\bar{F} \to L\bar{F}$ is a functor too by the discussion in 1.8. Now let be $\bar{F}: Fin^{op} \to Ban$ and $\bar{F}_1: Ban^{op} \to Ban$. Then we have

$$\underset{Ban}{Nat}(L\bar{F},\bar{F}_1)$$

$$= \underset{(.)\in Ban}{Nat}\left(\bar{F}(..) \underset{(..)\in Fin}{\hat{\otimes}} H(.,..), \bar{F}_1(.)\right)$$

$$= \underset{(..)\in Fin}{Nat}\left(\bar{F}(..), \underset{(.)\in Ban}{Nat}(H,(.,..), \bar{F}_1(.))\right) \text{ by}$$

an exponential law similar to 1.8.

$$= \underset{(..)\in Fin}{Nat}(\bar{F}(..), \bar{F}_1(..)) \text{ by the Yoneda lemma}$$

$$= \underset{Fin}{Nat}(\bar{F},\bar{F}_1|Fin).$$

This an adjointness relation, since its naturality (in $\bar{F}$ and $\bar{F}_1$) is implied by the naturality of the exponential law. qed.

2.10. A functor $\overline{F}$: $Ban^{op} \to Ban$ is said to be computable if $L(\overline{F}|Fin) = \overline{F}$.

Proposition: If $\overline{F}$: $Ban^{op} \to Ban$ is computable, then

$$\underset{Ban}{Nat}(\overline{F},\overline{F}_1) = \underset{End}{Nat}(\overline{F}|Fin,\overline{F}_1|Fin)$$

for all functors $\overline{F}_1$: $Ban^{op} \to Ban$.

Proof: $\underset{Ban}{Nat}(\overline{F},\overline{F}_1) = \underset{Ban}{Nat}(L(\overline{F}|Fin),\overline{F}_1)$

$$= \underset{Fin}{Nat}(\overline{F}|Fin,F_1|Fin).$$

2.11. Given $X \in Ban$ we consider the spectral family $\{X/M,\ M \subset X,\ X/M \in Fin\}$ in Fin, given by all quotients X/M of X over closed finite-codimensional subspaces M and canonical quotient maps $X/M \to X/M_1$ for $M_1 \supset M$. As we shall see later on $X \neq \underleftarrow{\lim}\{X/M,\ X/M \in Fin\}$ since $\underleftarrow{\lim}\{X/M, X/M \in Fin\} = X''$ (see 2.13. below).

Proposition: We have $L\overline{F}(X) = \underrightarrow{\lim}\{\overline{F}(X/M), X/M \in Fin\}$ for all $X \in Ban$ and $\overline{F}$: $Fin^{op} \to Ban$.

Proof: Write $L'\overline{F}(X) = \underrightarrow{\lim}\{\overline{F}(X/M),\ X/M \in Fin\}$, then $L'\overline{F}$ defines a contravariant functor $Ban^{op} \to Ban$: using the universal property of colimits it is readily seen L' is a functor too and is left adjoint to the restriction functor $\overline{F}_1 \to \overline{F}_1|Fin$. Since left adjoints are uniquely determined up to isometric natural isomorphisms, we have $L' = L$.

<u>2.12. Proposition: For any $\overline{F}$: $Ban^{op} \to Ban$ we have</u>

$$L\overline{F}(X) = \overline{F}(.) \mathop{\hat{\otimes}}_{(.)\in Fin} (X' \hat{\hat{\otimes}} .) = \overline{F}(.) \mathop{\hat{\otimes}}_{(.)\in Ban} (X' \hat{\hat{\otimes}} .),$$

<u>thus $L\overline{F}$ is always of type</u> Σ.

<u>Proof</u>: We wrote $L\overline{F}$ for $L(\overline{F}|Fin)$. Similarly as in the proof of 2.5. we have a canonical map $\overline{F}(.) \mathop{\hat{\otimes}}_{(.)\in Fin} (X' \hat{\hat{\otimes}} .) \to \overline{F}(.) \mathop{\hat{\otimes}}_{(.)\in Ban} (X' \hat{\hat{\otimes}} .)$ whose adjoint is the following isometric isomorphism, thus this map is isometric unto too:

$$(\overline{F}(.) \mathop{\hat{\otimes}}_{(.)\in Ban} (X' \hat{\hat{\otimes}} .))' = \mathop{Nat}_{Ban} (X' \hat{\hat{\otimes}} ., \overline{F}(.)') \text{ by 1.8.}$$

$$= \mathop{Nat}_{Fin} (X' \hat{\hat{\otimes}} ., \overline{F}(.)') \text{ by 2.3. and 2.7(b).}$$

$$= (\overline{F}(.) \mathop{\hat{\otimes}}_{(.)\in Fin} (X' \otimes .))'.$$

The last sentence of the proposition follows like 2.6 out 1.13e) qed.

<u>2.13. Examples</u>:

(a) $X \hat{\otimes} .'$ is computable:

$$L(X \hat{\otimes} .')(Y) = (X \hat{\otimes} .') \mathop{\hat{\otimes}}_{Ban} (Y' \hat{\hat{\otimes}} .) \text{ by 2.12}$$

$$= X \hat{\otimes} Y' \text{ by 1.13g).}$$

A special case of this result (for $X = I$) is:

$$Y' = L(.')(Y) = \lim_{\rightarrow} \{(Y/M)', Y/M \in Fin\} \text{ by 2.11}$$

$$= \lim_{\rightarrow} \{M^o, Y/M \in Fin\}$$

where M^o is the annihilator or the polar of M in Y', since $(Y/M)' = M^o$.

(b) $X \hat{\hat{\otimes}}\ .'$ is computable:

If M runs through all finite-codimensional closed subspaces of Y then $M^{\circ} = (Y/M)'$ runs through all finite dimensional subspaces of Y', and the spectral families coincide.

Thus $Y' = \lim\limits_{\rightarrow} \{M^{\circ},\ Y/M \in \mathrm{Fin}\}$

$= \lim\limits_{\rightarrow} \{E,\ E \subset Y',\ E \in \mathrm{Fin}\}$

$X \hat{\hat{\otimes}}\ Y' = X \hat{\hat{\otimes}}\ (\lim\limits_{\rightarrow} \{E,\ E \subset Y',\ E \in \mathrm{Fin}\})$

$= \lim\limits_{\rightarrow} \{X \hat{\hat{\otimes}}\ E,\ E \subset Y',\ E \in \mathrm{Fin}\}$ by 2.7(b)

$= \lim\limits_{\rightarrow} \{X \hat{\hat{\otimes}}\ (Y/M)',\ Y/M \in \mathrm{Fin}\}$

$= L(X \hat{\hat{\otimes}}.')(Y)$ by 2.11.

(c) $H(X,Y') = (X \hat{\otimes} Y)' = (Y \hat{\otimes} X)' = H(Y,X')$ shows that $'\ : \mathrm{Ban}^{op} \rightarrow \mathrm{Ban}$ is adjoint to itself on the right, thus ' transforms colimits into limits. Using this we conclude that $X'' = (X')' = (\lim\limits_{\rightarrow} \{E, E \subset X', E \in \mathrm{Fin}$

$= (\lim\limits_{\rightarrow} \{M^{\circ},\ X/M \in \mathrm{Fin}\})'$ by (b)

$= \lim\limits_{\leftarrow} \{(M^{\circ})',\ X/M \in \mathrm{Fin}\}$

$= \lim\limits_{\leftarrow} \{X/M,\ X/M \in \mathrm{Fin}\}$,

since $X/M \in \mathrm{Fin}$ is reflexive and $M^{\circ} = (X/M)'$. We will put this in a general framework later on (§8).

(d) We now give an example of a non-computable contravariant functor of type Σ.

Consider a Banach space X without the approximation property, the canonical map $s: Y' \hat{\otimes} X \to K(Y,X)$, given by $Y' \otimes X \to (Y' \to \langle y,y' \rangle x)$ and its canonical factorisation (compare 1.12):

$$\begin{array}{ccc} Y' \hat{\otimes} X & \xrightarrow{s} & K(X,X) \\ \text{coims} \downarrow & & \uparrow \text{ims} \\ N^1(Y,X) & \xrightarrow{s} & Y' \hat{\hat{\otimes}} X, \end{array}$$

where

$N^1(Y,X) = Y' \hat{\otimes} X / s^{-1}(0)$ is the space of all nuclear maps $Y \to X$. Then we know that $L(N^1(.,X)) = .' \hat{\otimes} X$ since $N^1(.,X)$ and $.' \hat{\otimes} X$ coincide on Fin and the latter functor is computable, but $N^1(X,X) \neq X' \hat{\otimes} X$ iff X has not the approximation property.

That this counterexample is again the typical one follows from the next proposition:

__2.14.__ __Proposition__: __If__ $F: Ban^{op} \to Ban$ __is any functor and__ X' __has the metric approximation property,__ __then__ $L\overline{F}(X) = \overline{F}_e(X)$.

Thus the contravariant computable functors $\underline{A}^{op} \to Ban$ are again exactly those of type Σ.

__Proof__: $L\overline{F}(X) = \overline{F}(.) \underset{(.)\in Ban}{\hat{\otimes}} (X' \hat{\hat{\otimes}} .)$ by 2.12

$= \overline{F}_e(X)$ by 1.13d).

2.15. How does this fit into abstract category theory? Denote by V the restriction functor $F \to F|Fin$ and by J the embedding functor $Fin \to Ban$, then $LF = Lan_J F$ is the left-hand Kan extension of F along J for F: $Fin \to Ban$. See MACLANE [12], chapter X. The tensor product of functors 1.6 is the coend, see loc. cit. chapter IX. Similarly most of the results in §4 below can be interpreted as Kan-extensions, e.g. 4.4 contains a left Kan-extension and 4.6 contains a right one.

§3. Duality of functors

3.1. Remark: The notion of duality of functors was introduced by MITJAGIN-SVARTS [16] and further studied by LEVIN [10], NEGREPONTIS [17], CIGLER [4], HERZ-PELLETIER [9], and by LINTON [11] and some Russian authors from a more abstract point of view.

Definition: For a functor $F: Ban \to Ban$ the dual functor $DF: Ban \to Ban$ to F is defined by $DF(X) = \underset{Ban}{Nat}(F, X\hat{\otimes}.)$, the action on morphisms is clearly given by $DF(f)(y) = (f\hat{\otimes}.)\circ y$ for $f: X \to Y$.

Remark: $D: (Ban^{Ban})^{op} \to Ban^{Ban}$ is a contravariant functor and is to itself adjoint at the right, i.e. the equality $\underset{Ban}{Nat}(F, DF_1) = \underset{Ban}{Nat}(F_1, DF)$ holds naturally in F and F_1.

Thus we have a distinguished natural transformation $\iota^F: F \to DDF$, corresponding to the 1_{DF} via $Nat(DF,DF) = Nat(F,DDF)$; in fact ι^F is the unit of the adjointness relation. F is said to be reflexive, if ι^F is isometric onto.

For further information see MITJAGIN SVARTS [16]; we are not interested here in the abstract properties of D, we want to compute DF for functors F of type Σ and to derive some results which will be useful in the theory of operator ideals later on.

We list some examples:

$D(H(X,.)) = X\hat{\otimes}.$ by the Yoneda lemma.

$D(X\hat{\otimes}.) = H(X,.)$.

3.2. Theorem: *If* $F\colon Ban \to Ban$ *is a functor of type* Σ, *then for any* $X \in Ban$ *we have* $DF(X) = \{f \in H(F(I),X)\colon i_X \circ f \in F(X')'\}$ *where* $i_X\colon X \to X''$ *is the canonical embedding and* $\|f\|_{DF(X)} = \|i_X \circ f\|_{F(X')'}$.

Proof: We should first loose some words on $F(X')'$: Since F is of type Σ we have $F(Y) = F(I) \otimes_\alpha Y$ where α is a reasonable crossnorm (1.9). Thus the canonical map $F(I) \hat{\otimes} Y \to F(I) \otimes_\alpha Y$ is contractive and is epimorphic (has dense image) and the adjoint map

$$F(Y)' = (F(I) \otimes_\alpha Y)' \to (F(I) \hat{\otimes} Y)' = H(F(I),Y') \text{ is therefore}$$

injective, i.e. each bounded linear functional on $F(I) \otimes_\alpha Y$ appears in $H(F(I),Y')$ and $F(Y)' = (F(I) \otimes_\alpha Y)'$ is the space of all $f \in H(F(I),Y')$ which define a continuous linear functional on $F(I) \otimes_\alpha Y$ by

$$\left\langle \sum_{i=1}^{n} a_i \otimes y_i, f \right\rangle = \sum_{i=1}^{n} \langle y_i, f(a_i) \rangle .$$

Now let us prove the theorem.

The map $j\colon DF(X) = Nat(F, X \hat{\otimes} .) \to H(F(I),X)$, defined by $j(\eta) = \eta_I$, is clearly contractive and injective since F is of type Σ: let be $\eta_I = 0$. For $z \in Z \in Ban$ set $\hat{z} \in H(I,Z)$, $\hat{z}(r) = r.z$.

The diagram

$$\begin{array}{ccc} F(I) & \xrightarrow{\eta_I} & X = X \hat{\otimes} I \\ F(z) \downarrow & & \downarrow 1_X \hat{\otimes} z \\ F(Z) & \xrightarrow{\eta_Z} & X \hat{\otimes} Z \end{array}$$

thus comutes, for all $a \in F(I)$ we have

$$\eta_Z F(\hat{z})\, a = (1_X \hat{\otimes}\, \hat{z})\, \eta_I(a) = 0.$$

Thus $\eta_Z\,(\sum_i F(\hat{z}_i)a_i) = 0$ for all $\Sigma a_i \otimes z_i \in F(I) \otimes Z$, the latter space is dense in $F(Z)$, so $\eta_Z = 0$ and since Z was arbitrary $\eta = 0$.

Take any $\eta \in \mathrm{Nat}(F, X \hat{\otimes} .)$. Then we assert that $i_X \circ \eta_I \in F(X')'$:

The diagram

$$\begin{array}{ccc} F(I) & \xrightarrow{\eta_I} & X \\ F(\hat{x}')\downarrow & & \downarrow 1_X \hat{\otimes}\, \hat{x}' \\ F(X') & \xrightarrow{\eta_{X'}} & X \hat{\otimes} X' \end{array}$$

commutes for all $x' \in X'$, where $\hat{x}' \in H(I,X')$. By $Tr: X \hat{\otimes} X' \to I$ let us denote the trace functional, corresponding to $i_X \in H(X,X'')=(X \hat{\otimes} X')'$, $Tr(x \otimes x') = \langle x,x'\rangle$. Then for all $a \in F(I)$ and $x' \in X'$ we have:

$$\begin{aligned} \langle \eta_I(a), x'\rangle &= Tr(\eta_I(a) \otimes x') \\ &= Tr \circ (1_X \hat{\otimes}\, \hat{x}') \circ y_I(a) \\ &= Tr \circ \eta_{X'} \circ F(\hat{x}')a; \end{aligned}$$

For $\sum_{i=1}^{n} a_i \otimes x_i' \in F(I) \otimes X'$ we compute:

$$\begin{aligned} \langle \sum_{i=1}^{n} a_i \otimes x_i', i_X \circ \eta_I\rangle &= \sum_{i=1}^{n} \langle x_i', i_X \circ \eta_I(a_i)\rangle \\ &= \sum_{i=1}^{n} \langle \eta_I(a_i), x_i'\rangle \\ &= \sum_{i=1}^{n} Tr \circ \eta_{X'} \circ F(\hat{x}_i')a_i \\ &= Tr \circ \eta_{X'} \circ \varepsilon^F_{X'}\,(\sum_{i=1}^{n} a_i \otimes x_i'), \end{aligned}$$

where $\varepsilon^F_{X'}: F(I) \otimes X' \to F(X')$ is the map of 1.9.

Thus

$$|\langle \sum_{i=1}^{n} a_i \otimes x_i', i_X \circ \eta_I \rangle| =$$

$$= |\mathrm{Tr} \circ \eta_{X'}, (\sum_{i=1}^{n} F(\hat{x}_i') a_i)|$$

$$\leq \|\mathrm{Tr}\| \; \|\eta_{X'}\| \; \|\sum_{i=1}^{n} F(\hat{x}_i') a_i\|_{F(X')},$$

i.e. $\| i_X \circ \eta_I \|_{F(X')'} \leq \|\eta_{X'}\| \leq \|\eta\|$.

Let us suppose conversely that we have $f \in H(F(I),X)$ with $i_X \circ f \in F(X')'$.

For any $Z \in \mathrm{Ban}$ we define

$$(\theta f)_Z : F(I) \otimes Z \to X \otimes Z$$

by $(\theta f)_Z (\Sigma a_i \otimes z_i) = \Sigma f(a_i) \otimes z_i$.

$$\| (\theta f)_Z (\sum_{i=1}^{n} a_i \otimes z_i)\|_{X \hat{\otimes} Z}$$

$$= \|\sum_{i=1}^{n} z_i \otimes f(a_i)\|_{Z \hat{\otimes} X}$$

$$= \sup_{h \in H(Z,X'), \|h\| \leq 1} |\langle \sum_{i=1}^{n} z_i \otimes f(a_i), h\rangle|$$

$$= \sup_{h \in OH(Z,X')} |\sum_{i=1}^{n} \langle f(a_i), h(z_i)\rangle|$$

$$= \sup_{h \in OH(Z,X')} |\sum_{i=1}^{n} \langle h(z_i), i_X \circ f(a_i)\rangle|$$

$$= \sup_{h \in OH(Z,H')} |\langle \sum_{i=1}^{n} h(z_i) \otimes a_i, i_X \circ f\rangle|$$

$$\leq \sup_{h \in OH(Z,X')} \|\sum_{i=1}^{n} F(\widehat{h(z_i)}) a_i\|_{F(X')} \| i_X \circ f \|_{F(X')'}$$

$$= \sup_{h \in OH(Z,X')} \| F(h) \sum_{i=1}^{n} F(\hat{z}_i) a_i\|_{F(X')} \| i_X \circ f\|_{F(X')'}$$

$$\leq \|\sum_{i=1}^{n} F(\hat{z}_i) a_i\|_{F(Z)} \| i_X \circ f\|_{F(X')'}.$$

Thus $(\theta f)_Z$ extends to a continuous map $F(Z) \to X \hat{\otimes} Z$ with $\| (\theta f)_Z \| \leq \| i_X \circ f \|_{F(X')'}$. By the naturality of the counit ε^F it is very easily seen that $((\theta f)_Z)$ is a natural transformation $F \to X \hat{\otimes} .$, clearly we have $(\theta f)_I = f$, and since the map $j: DF \to H(F(I),.)$ above is easily seen to be natural we are done. qed.

<u>Remark</u>: HERZ-PELLETIER [9] had that result for computable functors, see their Corollary 2.9.

As a special case we find that

$$D(X \hat{\hat{\otimes}} .)(Y) = \{f: X \to Y: i_Y \circ f \in (X \hat{\hat{\otimes}} Y')'\}$$

$$= I_1(X,Y), \text{ the space of integral operators } X \to Y, \text{ see}$$

GROTHENDIECK [8].

This result can be found in CIGLER [5], page 151.

<u>3.3.</u> <u>Theorem</u> (HERZ-PELLETIER, [9] theorem 1.9):

<u>If</u> F: Ban $\to$ Ban <u>is computable</u>, <u>then</u>

$$DF(X') = F(X)'$$

<u>Proof</u>: This proof is much simpler than the original one:

$$\begin{aligned} DF(X') &= \operatorname{Nat}_{Ban} (F, X' \hat{\otimes} .) \\ &= \operatorname{Nat}_{(.) \in Ban} \left(H(..,.) \hat{\otimes}_{(..) \in Fin} F(..), X' \hat{\otimes} .\right) \text{ by 2.4} \\ &= \operatorname{Nat}_{(..) \in Fin} \left(F(..), \operatorname{Nat}_{(.) \in Ban} (H(..,.), X' \hat{\otimes} .)\right) \text{ by 1.8} \\ &= \operatorname{Nat}_{(..) \in Fin} (F(..), X' \hat{\otimes} ..) \text{ by the Yoneda lemma} \\ &= \operatorname{Nat}_{(..) \in Fin} (F(..), (X \hat{\hat{\otimes}} ..')'), \text{ since } (..) \in Fin \text{ and} \end{aligned}$$

for $E \in Fin$ we have $H(E,X)' = (E' \hat{\hat{\otimes}} X)' = E \hat{\otimes} X'$.

$$= ((..' \hat{\hat{\otimes}} X) \hat{\otimes}_{(..)\in Fin} F(..))' \text{ by 1.8.}$$

$$= F(X)' \text{ since F is computable.} \qquad \text{qed.}$$

3.4. If we define $D'F(X) = \underset{\underline{A}}{\text{Nat}} (F, X \hat{\otimes} .)$, then we have for any functor F of type Σ that $D'F(X') = F(X)'$ by 2.8 and 3.3, whether $X \in \underline{A}$ or not. A related result is the following:

Theorem: If F: Ban → Ban is of type Σ and X' has the metric approximation property, then $DF(X') = F(X)'$.

Proof: If X' has the metric approximation property, then it is well known that X has it too. Thus we have

$$F(X) = (.' \hat{\hat{\otimes}} X) \hat{\otimes}_{(.)\in Ban} F(.) \quad \text{by 1.13c).}$$

Then
$$F(X)' = [(.' \hat{\hat{\otimes}} X) \hat{\otimes}_{(.)\in Ban} F(.)]'$$

$$= \underset{Ban}{\text{Nat}} (F(.), (.' \hat{\hat{\otimes}} X)') \text{ by 1.8}$$

$$= \underset{Ban}{\text{Nat}} (F(.), (.' \hat{\hat{\otimes}} X)'_e) \text{ by 1.11}$$

$$= \underset{Ban}{\text{Nat}} (F(.), X' \hat{\otimes} .) = DF(X').$$

The last equality holds since $X' \in \underline{A}$ by GROTHENDIECK [7], page 181, §5, No. 2, Prop. 40, Corr.1. qed

3.5. Since we will need it later we introduce now an analogous notion of duality for contravariant functors.

<u>Definition</u>: Let $\overline{F}$: $Ban^{op} \to Ban$ be a contravariant functor. Define $D\overline{F}$: $Ban^{op} \to Ban$ by

$$D\overline{F}(X) = \operatorname{Nat}_{Ban} (\overline{F}, X' \hat{\otimes} .'),$$

$$D\overline{F}(f)(\eta) = (f' \hat{\otimes} .) \circ \eta.$$

Clearly $D\overline{F}$ is again a contravariant functor, D itself is a functor and is adjoint to itself at the right, i.e. $\operatorname{Nat}(\overline{F}, D\overline{F}_1) = \operatorname{Nat}(\overline{F}_1, D\overline{F})$ holds naturally in $\overline{F}$ and $\overline{F}_1$. That can be proved analogously as the same relation for covariant duality is proved in MITJAGIN-SVARTS [16]. We can introduce even a notion of reflexivity. But we will not need any of these developments later on. We conclude with the most elementary examples: $D(H(.,A)) = A' \hat{\otimes} .'$, $D(A \hat{\otimes} .') = H(.,A')$, which can be computed fairly easily.

<u>3.6. Theorem</u>: <u>If</u> $\overline{F}$: $Ban^{op} \to Ban$ <u>is of type</u> Σ, <u>then</u>

$$\operatorname{Nat} (\overline{F}, X \hat{\otimes} .') = \{f \in H(\overline{F}(I), X) : i_X \circ f \in \overline{F}(X)'\}$$

<u>with</u> $\|f\|_{\operatorname{Nat}(\overline{F}, X \hat{\otimes} .')} = \|i_X \circ f\|_{\overline{F}(X)'}$.

Thus $D\overline{F}(X) = \operatorname{Nat}(\overline{F}, X' \hat{\otimes} .')$

$$= \{f \in H(\overline{F}(I), X') : i_{X'} \circ f \in \overline{F}(X')'\}$$

with $\|f\|_{D\overline{F}(X)} = \|i_{X'} \circ f\|_{\overline{F}(X')'}$.

<u>Proof</u>: By 1.10 $\overline{F}(X) = \overline{F}(I) \otimes_\alpha X'$, where α is a reasonable crossnorm. An argument similar to that in the proof of theorem 3.2. shows that $\overline{F}(X)' = (\overline{F}(I) \otimes_\alpha X')' \subset H(\overline{F}(I), X'')$.

The rest of the proof is the same as for the theorem 3.2 with the obvious changes and we do not repeat it. qed.

<u>3.7. Proposition: If $\overline{F}$: $Ban^{op} \to Ban$ is computable (2.10), then</u>

$D\overline{F}(X') = \mathrm{Nat}(\overline{F}, X'' \hat{\otimes} .') = \overline{F}(X)'$.

<u>Proof</u>: $\underset{Ban}{\mathrm{Nat}}(\overline{F}, X'' \hat{\otimes} .')$

$$= \underset{(.)\in Ban}{\mathrm{Nat}} \left(\overline{F}(..) \underset{(..)\in Fin}{\hat{\otimes}} H(.,..), X'' \hat{\otimes} .'\right) \text{ by 2.9}$$

$$= \underset{(..)\in Fin}{\mathrm{Nat}} \left(\overline{F}(..), \underset{(.)\in Ban}{\mathrm{Nat}} (H(.,..), X'' \hat{\otimes} .')\right) \text{ by 1.8.}$$

$$= \underset{(..)\in Fin}{\mathrm{Nat}} (\overline{F}(..), X'' \hat{\otimes} ..') \text{ by the Yoneda lemma}$$

$$= \underset{(..)\in Fin}{\mathrm{Nat}} (\overline{F}(..), (X' \hat{\hat{\otimes}} ..)'), \text{since } (..) \in Fin$$

and for $E \in Fin$ we have $X'' \hat{\otimes} E' = (X' \hat{\hat{\otimes}} E)'$.

$$= \left(\overline{F}(..) \underset{(..)\in Fin}{\hat{\otimes}} (X' \otimes ..)\right)' \text{ by 1.8}$$

$$= \overline{F}(X)', \text{since } \overline{F} \text{ is computable.}$$

<u>3.8.</u> Had we considered $D\overline{F}$ to be defined by $D\overline{F}(X) = \underset{\underline{A}'}{\mathrm{Nat}}(\overline{F}, X' \hat{\otimes} .')$, where $\underline{A}'$ is the full subcategory of those Banach spaces X such that X' has the metric approximation property, then by 2.14 any functor $\overline{F}$ of type Σ is computable on $\underline{A}'$ and we would have $D\overline{F}(X') = \overline{F}(X)'$ for all functors of type Σ. A related result is the following.

<u>3.9. Proposition: If $\overline{F}$: $Ban^{op} \to Ban$ is of type Σ and X" has the metric approximation property, then</u> $D\overline{F}(X') = \underset{Ban}{\mathrm{Nat}}(\overline{F}, X'' \hat{\otimes} .') = \overline{F}(X)'$.

Proof: X' has the metric approximation property too, since X" has it, thus $\bar{F}(X) = \bar{F}(.) \hat{\otimes}_{(.)\in Ban} (X' \hat{\hat{\otimes}} .)$ by 1.13d). So we have:

$$\bar{F}(X)' = (\bar{F}(.) \hat{\otimes}_{(.)\in Ban} (X' \hat{\hat{\otimes}} .))'$$

$$= \mathrm{Nat}_{Ban} (\bar{F}(.), (X' \hat{\hat{\otimes}} .)') \text{ by 1.8}$$

$$= \mathrm{Nat}_{Ban} (\bar{F}(.), (X' \hat{\hat{\otimes}} .)'_e) \text{ by 1.11}$$

$$= \mathrm{Nat}_{Ban} (\bar{F}(.), X'' \hat{\otimes} .') = D\bar{F}(X'),$$

by the result cited in the proof of theorem 3.4, since $X'' \in \underline{A}$. qed.

3.10. We will treat now one of the main situations in which we will need duality of functors.

Definition: A contractive morphism $f: X \to Y$ is said to be a weak retract, if there exists $h: X' \to Y'$, $\|h\| \leq 1$, such that $f' \circ h = 1_{X'}$.
f is said to be a weak section, if there exists $g: X' \to Y'$, $\|g\| \leq 1$, such that $g \circ f' = 1_{Y'}$.

f is a weak retract iff it is isometric and

$0 \to f'^{-1}(0) = f(X)^0 \to Y' \to X' \to 0$ is a splitting short exact sequence in Ban_1.

f is a weak section iff, it is a quotient map and

$0 \to Y' \to X' \to X'/f'(Y') = (f^{-1}(0))' \to 0$

is a splitting short exact sequence in Ban_1. The equality $(i_X)' \circ i_{X'} = 1_{X'}$ shows that i_X is a weak retract and the equation $(i_{X'})' \circ i_{X''} = ((i_X)' \circ i_{X'})' = (1_{X'})' = 1_{X''}$ shows that $i_{X'}: X''' \to X'$ is a weak section.

<u>3.11.</u> <u>Lemma</u>: (HERZ-PELLETIER [9], lemma 2.7)

$f: X \to Y$, $\|f\| \leq 1$ <u>is a weak retract if and only if</u>

$Z \hat{\otimes} f: Z \hat{\otimes} f: Z \hat{\otimes} X \to Z \hat{\otimes} Y$ <u>is isometric for all</u> $Z \in$ Ban.

<u>Proof</u>: $Z \hat{\otimes} f$ is isometric iff $(Z \hat{\otimes} f)' = H(Z,f')$ is a quotient map, moreover $H(Z,f')(OH(Z,Y')) = OH(Z,X')$ since $H(Z,f')$ is $\sigma(H(Z,Y'), Z \otimes Y)$ continuous by a compactness argument. Choose h to be a preimage of $1_{X'}$ under $H(X',f'): H(X',Y') \to H(X',X')$ with norm ≤ 1. If f is a weak retract, then

$$H(Z,f')\,(h \circ g) = f' \circ h \circ g = g, \quad \|h \circ g\| \leq \|g\|$$

for $g \in H(Z,Y')$ shows that $H(Z,f')$ is a quotient map for all Z. qed.

<u>3.12.</u> <u>Proposition</u>: <u>A covariant computable functor</u> F <u>transforms weak retracts into weak retracts and weak sections into weak sections</u>. <u>A contravariant computable functor</u> $\overline{F}$ <u>transforms weak retracts into weak sections and weak sections into weak retracts</u>.

<u>Proof</u>: Let $\overline{F}$: Ban^{op} → Ban, F: Ban → Ban be functors. Then $f' \circ h = 1_{X'}$ implies $F(f)' \circ DF(h) = DF(f') \circ DF(h)$ by 3.3.

$$= DF(f' \circ h)$$

$$= DF(1_{X'}) = 1_{DF(X')} = 1_{F(X)'}.$$

$$D\overline{F}(h) \circ \overline{F}(f)' = D\overline{F}(h) \circ D\overline{F}(f') \text{ by 3.7.}$$

$$= D\overline{F}(f' \circ h)$$

$$= D\overline{F}(1_{X'}) = 1_{D\overline{F}(X')} = 1_{\overline{F}(X)'}.$$

$h \circ f' = 1_Y$ implies in turn

$$DF(h) \circ F(f)' = DF(h)\, DF(f') \text{ by 3.3.}$$

$$= DF(h \circ f') = DF(1_{X'})$$

$$= 1_{DF(X')} = 1_{F(X)'}.$$

$$\overline{F}(f)' \circ D\overline{F}(h) = D\overline{F}(f') \circ D\overline{F}(h) \text{ by 3.7.}$$

$$= D\overline{F}(h \circ f') = D\overline{F}(1_{X'})$$

$$= 1_{D\overline{F}(X')} = 1_{\overline{F}(X)'}.$$

qed.

3.13. Proposition: (HERZ-PELLETIER [9], 2.8)

If F is of type Σ and $f: X \to Y$ is a weak retract, then the following diagram is a pullback in Ban_1:

$$\begin{array}{ccc} Nat(F, X \hat{\otimes} .) & \xrightarrow{j} & H(F(I),X) \\ DF(f) \downarrow & & \downarrow H(F(I),f) \\ Nat(F, Y \hat{\otimes} .) & \xrightarrow{j} & H(F(I),Y), \end{array}$$

where $j(\eta) = \eta_I$.

Proof: We give a shorter proof. J is injective since F is of type Σ (see the beginning of the proof of theorem 3.2). H(F(I),f) and DF(f) are isometries. Since H(F(I),f) is isometric and j is injective, the pullback of the half-diagram is just $j^{-1}(H(F(I),f)H(F(I),X))$

$= \{\eta \in Nat(F, Y \hat{\otimes} .): \eta_I(F(I)) \subset X\}$

$= Nat\ (F, X \hat{\otimes} .)$ since DF(f) too is isometric. qed.

A similar result holds for contravariant functors.

3.14. Corollary: If F is of type Σ then we have for $f \in H(F(I),X)$, using theorem 3.2: $f \in DF(X)$ iff $i_X \circ f \in DF(X'')$, and

$$\|f\|_{DF(X)} = \|i_X \circ f\|_{DF(X'')}.$$

This is more general than 2.9 of HERZ-PELLETIER [9], since theorem 3.2. is more powerful.

For the case $F = X \hat{\hat{\otimes}}$. this boils down to the well known result of GROTHENDIECK [8]. $f: X \to Y$ is integral iff $i_X \circ f$ is integral and their integral norms coincide.

<u>3.15</u> <u>Example</u>: Let X be without and Y be with approximation property. Then there is no weak retract $X \to Y$.

<u>Proof</u>: Suppose $f: X \to Y$ is a weak retract.

Consider

$$\begin{array}{ccc} X \hat{\otimes} X' & \xrightarrow{f \hat{\otimes} X'} & Y \hat{\otimes} X' \\ s \downarrow & & \downarrow s \\ L(X',X') & \xrightarrow{L(f',X')} & L(Y',X'), \end{array}$$

where both horizontal maps are isometric and the right hand side s is injective, since Y has the approximation property (see 1.12). So the left hand side s should be injective too, thus X should have the approximation property too (see GROTHENDIECK [7], p. 164), a contradiction. qed.

Thus the canonical embedding $X \to C(OX')$ into the space of continuous functions on the dual ball with weak-*-topology is in general no weak retract.

§4. Tensor products and operator ideals

4.1. In this chapter we want to give definitions of tensor products and operator ideals in terms of category theory and we want to reveal some relationships between them, which are mainly due to GROTHENDIECK [8]. In this section G is always supposed to be a contra-covariant bifunctor $Ban^{op} \times Ban \to Ban$ and M is a co-covariant one: $Ban \times Ban \to Ban$, which are supposed to satisfy $G(I,I) = I$ and $M(I,I) = I$ in the second half of this section (from 4.7 onwards).

4.2. Proposition: $$\operatorname{Nat}_{Ban^{op}\times Ban}(H(.,X) \hat{\otimes} H(Y,..),G) = G(X,Y)$$

$$\operatorname{Nat}_{Ban \times Ban}(H(X,.) \hat{\otimes} H(Y,.),M) = M(X,Y)$$

hold naturally in $X,Y \in Ban$ and in G,M.

Proof: This is just a special case of the Yoneda lemma, if one considers multilinear categories. We will however sketch an elementary proof of the first relation, the second being similar.

Define $\operatorname{Nat}(\ldots) \underset{\theta}{\overset{\psi}{\rightleftarrows}} G(X,Y)$ by

$\psi(\varphi) = \varphi_{XY}(1_X \otimes 1_Y) \in G(X,Y)$, $\varphi \in \operatorname{Nat}(\ldots)$

$(\theta g)_{Z_1Z_2}(f \otimes h) = G(f,h)g$, $g \in G(X,Y)$ and $f \otimes h \in H(Z_1,X) \hat{\otimes} H(Y_1 Z_2)$.

Routine computation shows that ψ,θ are contractive and linear and that e.g. ψ is natural and that $\theta = \psi^{-1}$ holds. qed.

<u>4.3.</u> <u>Proposition</u> $\underset{\mathrm{Ban}^{op}\times \mathrm{Ban}}{\mathrm{Nat}}(.'\hat{\otimes} A\hat{\otimes}..,G) = H(A,G(I,I))$,

$$\underset{\mathrm{Ban}\times \mathrm{Ban}}{\mathrm{Nat}}(.\hat{\otimes} A\hat{\otimes}..,M) = H(A,M(I,I))$$

<u>hold naturally in</u> $A \in \mathrm{Ban}$ <u>and in</u> M,G.

<u>Proof</u>: Again for the first equation only; the result is a bilinear version of 1.9, 1.10.

Define $\mathrm{Nat}\,(\ldots) \underset{\theta}{\overset{\psi}{\rightleftarrows}} H(A,G(I,I))$ by

$\psi(\varphi) = \varphi_{II} \in H(A,G(I,I))$, $\varphi \in \mathrm{Nat}\,(\ldots)$,

$(\theta f)_{XY}\,(x' \otimes a \otimes y) = G(x', \hat{y})\, f(a)$,

$f \in H(A,G(I,I))$, $x' \otimes a \otimes y \in X' \hat{\otimes} A \hat{\otimes} Y$.

Again it is routine computation to check up that ψ,θ are linear, contractive, that e.g. ψ is natural and that $\theta = \psi^{-1}$. qed.

<u>4.4.</u> We can interpret 4.3. as an adjunction: the "free" functor $A \mapsto (.'\hat{\otimes} A\hat{\otimes}\,..)$ is left adjoint to the forgetful functor $G \mapsto G(I,I)$. The unit of this adjunction is trivial, the counit is the map $\varepsilon^G_{XY}\colon X' \hat{\otimes} G(I,I) \hat{\otimes} Y \to G(X,Y)$, given by $\varepsilon^G_{XY}(\Sigma\, x_i' \otimes a_i \otimes y_i) = \Sigma\, G(x_i', \hat{y}_i)\, a_i$; it is contractive and natural in X,Y and G.

In order to break down this situation to known results let us denote by

$\varepsilon^{G(.,Z)}_X : X' \hat{\otimes} G(I,Z) \to G(X,Z)$ and

$\varepsilon^{G(Z,.)}_Y : G(Z,I) \hat{\otimes} Y \to G(Z,Y)$ the counits of the partial functors of G, introduced in 1.10 and 1.9 respectively. Then for all $X, Y \in \mathrm{Ban}$ the following diagram comutes, since that is clearly true on the dense subspace $X' \otimes G(I,I) \otimes Y$:

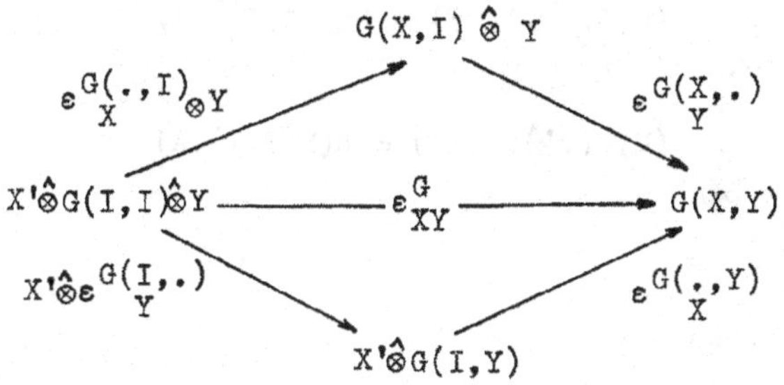

<u>Theorem</u>: <u>For all</u> $X, Y \in$ Ban <u>and for all</u> G <u>the map</u> ε_{XY}^G <u>restricted to</u> $X' \otimes G(I,I) \otimes Y$ <u>is injective and</u> $||.||_{G(X,Y)}$ <u>induces a reasonable norm</u> α <u>on</u> $X' \otimes G(I,I) \otimes Y$ (<u>reasonable means here</u>

$$\| \cdot \|_{X'\check{\otimes} G(I,I)\check{\otimes} Y} \leq \alpha \leq \| \cdot \|_{X'\hat{\otimes} G(I,I)\hat{\otimes} Y})$$

<u>The same is true for functors</u> M <u>with the analogous maps</u>.

<u>Proof</u>: All four inclined maps in the diagram above satisfy this by 1.9, 1.10.

The map ε_{XY}^M: $X\hat{\otimes}M(I,I) \hat{\otimes} Y \to M(X,Y)$ is given by

$$\varepsilon_{XY}^M \left(\Sigma\, x_i \otimes m_i \otimes y_i\right) = \Sigma\, M(\hat{x}_i, \hat{y}_i)\, m_i. \qquad \text{qed}$$

<u>Definition</u>: G respectively M is said to be a bifunctor of type Σ, if for all X, $Y \in$ Ban the maps ε_{XY}^G respectively ε_{XY}^M have dense image in G(X,Y) respectively M(X,Y).

In general we denote by $G_e(X,Y)$ respectively $M_e(X,Y)$ the closure of the image of ε_{XY}^G respectively ε_{XY}^M in G(X,Y) respectively M(X,Y); that defines a partial functor which we call the type -Σ-part or essential part of G respectively M.

Thus bifunctors of type Σ are essentially given by tensor products.

4.5. Proposition: $\underset{\mathrm{Ban}^{op}\times \mathrm{Ban}}{\mathrm{Nat}} (G, H(.\hat{\otimes}.', A)) = H(G(I,I),A)$,

$\underset{\mathrm{Ban}\times \mathrm{Ban}}{\mathrm{Nat}} (M, H(.'\hat{\otimes}.', A)) = H(M(I,I),A)$

hold naturally in $A \in \mathrm{Ban}$, G, M.

Proof for the first relation only:

Define $\mathrm{Nat}\,(\ldots) \underset{\theta}{\overset{\psi}{\rightleftarrows}} H(G(I,I),A)$ by

$\psi(\varphi) = \varphi_{I,I}$, $\varphi \in \mathrm{Nat}(\ldots)$.

$(\theta f)_{XY}(g)\ (x \otimes y') = f \circ G(\hat{x}, y')g$,

$f \in H(G(I,I),A)$, $g \in G(X,Y)$, $x \otimes y' \in X \hat{\otimes} Y'$.

It is a routine matter to prove that ψ, θ are contractive, that ψ is natural and that $\psi = \theta^{-1}$ holds.

4.6. This result too is an adjunction: the functer $A \to H(.\hat{\otimes}.', A)$ is right adjoint to the forgetful functor $G \to G(I,I)$. The counit of this adjunction is the map

$\varphi^G_{XY} : G(X,Y) \to H(X \hat{\otimes} Y', G(I,I))$, given by

$(\varphi^G_{XY}(g))\ (x \otimes y') = G(\hat{x}, y')\ g$, $g \in G(X,Y)$. φ^G_{XY} is contractive and natural in G and X,Y. The counit of the adjunction for M is

$\varphi^M_{XY}: M(X,Y) \to H(X' \hat{\otimes} Y', M(I,I))$, given by $\varphi^M_{XY}(m)\ (x' \otimes y') = M(x',y')m$, $m \in M(X,Y)$.

Definition: G,M are said to be total bifunctors if for all X, Y the maps φ^G_{XY}, φ^M_{XY} are injective. $G(M)$ is total iff maps of the form $G(\hat{x},y')$, $x \in X$, $y' \in Y'$ ($M(x',y')$, $x' \in X'$, $y' \in Y'$) separate points on $G(X,Y)$ ($M(X,Y)$) for all X and $Y \in \mathrm{Ban}$.

4.7. From now on to the end of this section we suppose that G and M satisfy the condition $G(I,I) = I$ and $M(I,I)$.

Definition: A tensor product is a co-covariant bifunctor $M: \text{Ban} \times \text{Ban} \to \text{Ban}$ with $M(I,I) = I$ of type Σ.

This definition is justified by 4.4, since $X \otimes Y$ is dense in $M(X,Y)$, $\| \cdot \|_{M(X,Y)}$ is a reasonable norm, and the tensor product is bifunctorial, i.e. the map $f \otimes g: X \otimes Y \to X_1 \otimes Y_1$ extends to $M(f,g): M(X,Y) \to M(X,Y)$ and $\|f \otimes g\| = \| M(f,g) \| \leq \|f\|\|g\|$ for all $f \in H(X,X_1)$ and $g \in H(Y,Y_1)$. We will write $X \otimes_M Y$ sometimes for $M(X,Y)$.

A tensor product M is said to be computable if all partial functors $M(.,Y), M(X,.)$ are computable. Equivalent are the conditions

$$M(X,Y) = H(.,X) \underset{(.)\in Fin}{\hat{\otimes}} (H(..,Y) \underset{(..)\in Fin}{\hat{\otimes}} M(.,..))$$

$$= (H(.,X) \underset{(.)\in Fin}{\hat{\otimes}} M(.,..)) \underset{(..)\in Fin}{\hat{\otimes}} H(..,Y),$$

where we were a little unprecise on the order. We could change brackets, since the tensor product of functors is associative (1.13f). Thus we have for a computable tensor product (2.2):

$$M(X,Y) = \underset{\to}{\lim} \{\underset{\to}{\lim} \{M(E,F),\ F \subset Y,\ F \in Fin\},\ E \subset X,\ E \in Fin\}$$

$$= \underset{\to}{\lim} \{\underset{\to}{\lim} \{M(E,F),\ E \subset X,\ E \in Fin\},\ F \subset Y,\ F \in Fin\}$$

$$= \underset{\to}{\lim} \{M(E,F),\ E \subset X,\ F \subset Y,\ E, F \in Fin\},$$

the change of order of the colimits is due to the associativity of the tensorproduct of functors. The following theorem is clear from that.

<u>Theorem</u>: <u>Computable tensor products correspond exactly to the</u> <u>$\otimes$-norms of</u> GROTHENDIECK [8].

<u>4.8.</u> Given a contra-covariant bifunctor G with $G(I,I) = I$, then the canonical map $\varphi^G_{XY}: G(X,Y) \to H(X \hat{\otimes} Y',I)$ (4.6) actually takes its image in $H(X,Y'')$ and is given by

$\langle Y', \varphi^G_{XY}(g)(x)\rangle = G(\hat{x},y')\, g \in I,\; g \in G(X,Y)$. Since φ^G is natural and contractive, the action of the bifunctor $H(.,..'')$ coincides with that of G if we consider $G(X,Y)$ as a (non-closed) subspace of $H(X,Y'')$ via φ^G_{XY}; the norm of $G(X,Y)$ is greater than that of $H(X,Y'')$, we express this fact by saying that $G(X,Y)$ is contractively contained in $H(X,Y'')$, or that G is a subfunctor of $H(.,..'')$ (in contrary a partial functor is an isometrically contained functor, 1.9, 1.10).

Now via some canonical map we have:

$$X' \otimes Y \subset G(X,Y) \subset H(X,Y'').$$

To know that all these inclusions are well defined we should check up that

$$\begin{array}{ccc} X' \hat{\otimes} Y & \xrightarrow{\varepsilon^G_{XY}} & G(X,Y) \\ & \searrow^{H(X,i_Y)\circ S} & \downarrow \varphi^G_{XY} \\ & & H(X,Y'') \end{array}$$

commutes

where $X' \hat{\otimes} Y \to H(X,Y'')$ is the canonically given map $x' \otimes y \mapsto (x \mapsto i_Y(\langle x,x'\rangle y))$. But this is rather trivial.

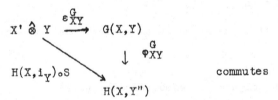

Since $\|\cdot\|_{G(X,Y)}$ induces a reasonable crossnorm on $X' \otimes Y$ we have

$X' \subset G(X,I) \subset H(X,I) = X'$,

$Y \subset G(I,Y) \subset H(I,Y'') = Y''$, where the first inclusions are isometrical. Thus $G(X,I) = X'$ for all X, but the covariant part does not behave as well; we should distinguish two cases:

<u>Definition</u>: A total bifunctor G with $G(I,I) = I$ is said to be of type (I), if $G(I,Y) = Y$ holds for any Y via the above inclusions. If $G(I,Y) = Y''$ for all Y, then G is said to be of type (II)

<u>Remark</u>: There is a total bifunctor G with $G(I,I) = I$ which is neither of type (I) nor of type (II).

Since we can factor φ^G_{XY} as

$G(X,Y) \to H(X,G(I,Y)) \to H(X,Y'')$, where the first map is given by $g \mapsto (x \mapsto G(\hat{x},Y)g)$ for $g \in G(X,Y)$, the canonical map φ^G_{XY} actually takes its values in $H(X,Y)$ if G is of type (I), and thus the expression

$$\langle \varphi^G_{XY}(g)(x),y' \rangle = G(\hat{x},y')g, \quad g \in G(X,Y) \text{ is weak-*-continuous}$$

and well defined.

<u>4.9.</u> <u>Definition</u>: A bifunctor A of type (I) is called an operator ideal. To justify this definition we will show that it coincides with the usual notion of a Banach operator ideal (see PIETSCH [18], [19], or GORDON-LEWIS-RETHERFORD [6] for a quick account and examples):

A class A of bounded linear operators between Banach spaces is a Banach operator ideal, if its components $A(X,Y) = A \cap H(X,Y)$ are linear subspaces of $H(X,Y)$, which are Banach spaces under a norm $\|.\|_A$ and satisfy the following conditions

(i) $x' \in X'$, $y \in Y$ implies $x' \otimes y \in A(X,Y)$
and $\|x' \otimes y\|_A = \|x'\| \; \|y\|$.

(ii) $f \in H(X_1,X)$, $g \in A(X,Y)$, $h \in H(Y,Y_1)$ implies

$$h \circ g \circ f \in A(X_1,Y_1) \text{ and } \|h \circ g \circ f\|_A \leq \|h\| \; \|g\|_A \; \|f\|.$$

Thus clearly each Banach operator ideal in the usual sense becomes a bifunctor of type (I) by putting $A(f,h)\, g = h \circ g \circ f$.

Conversely each bifunctor Λ of type (I) is a Banach operator ideal, condition (ii) being subsumed in the functorial property:

$$g \in \Lambda(X,Y),\ f \in H(X_1,X),\ h \in H(Y,Y_1), \text{ then } h \circ g \circ f = H(f,h)\, \varphi^{\Lambda}_{XY}(g)$$

$$= \varphi^{\Lambda}_{X_1Y_1}\, \Lambda(f,h)g$$

$$\|h \circ g \circ\|_{\Lambda} = \|\Lambda(f,h)g\|_{\Lambda\,(X_1,Y_1)} \leq \|f\| \; \|h\| \; \|g\|_{\Lambda\,(X,Y)},$$

where we identified g and $\varepsilon^{\Lambda}_{XY}(g)$ for short. We could collect all this in the following

Theorem: The Banach operator ideals in the usual sense are exactly the bifunctors of type (I) on Ban.

4.10 If G is a bifunctor of type (II), then we have an associated bifunctor $G^{(I)}$ of type (I), given by

$$G^{(I)}(X,Y) = \{f \in H(X,Y) : i_{\psi} \circ f \in G(X,Y) \text{ via } \varphi^G\}.$$

with the norm $\|f\|_{G(I)} = \|i_{\psi} \circ f\|_G$, or:

<u>Lemma</u>: $G^{(I)}(X,Y)$ <u>is the pullback of the diagram</u>

$$\begin{array}{ccc} & & H(X,Y) \\ & & \downarrow H(X,i_Y) \\ G(X,Y) & \xrightarrow{\varphi^G_{XY}} & H(X,Y'') \end{array} .$$

<u>Proof</u>: $H(X,i_Y)$ is isometric, φ^G_{XY} is injective, thus the pullback of this diagram is

$$(\varphi^G_{XY})^{-1} \;(H(X,i_Y)\;(H(X,Y))) =$$

$$\{g \in G(X,Y):\ \varphi^G_{XY} g \in H(X,i_Y)\;(H(X,Y))\}$$

$$= \;G^{(I)}(X,Y).$$

qed.

Since φ^G_{XY} is natural in X,Y and $H(X,i_Y)$ is natural in X,Y, and since $G^{(I)}(X,Y)$ is the pull back of these two maps, $G^{(I)}$ is a bifunctor, a partial functor of G, all values of its elements under φ^G_{XY} lie in H(X,Y), thus $G^{(I)}$ is of type (I).

Since φ^G is natural in G too, the map $G \mapsto G^{(I)}$ is a functor too, which assigns bifunctors of type (I) to total bifunctors G with G(I,I) = I. This functor is right adjoint to the embedding of bifunctors of type (I), i.e. $\mathrm{Nat}_{Ban^{op}\times Ban}(\Lambda,G) = \mathrm{Nat}_{Ban^{op}\times Ban}(\Lambda,G^{(I)})$ holds naturally in Λ of type (I) and in total G with G(I,I) = I, by the universal property of the pullback, or by a routine computation.

4.11. Proposition: Given a tensor product M, then

DM is an operator ideal, where

$DM(X,Y) := D(M(X,.))(Y)$. (3.1)

Furthermore we have

$DM = (M(.,..')')^{(I)}$.

Proof: $DM = (M(.,..')')^{(I)}$ holds by theorem 3.2.

$X \hat{\otimes} Y' \xrightarrow{\varepsilon^M_{XY}} M(X,Y')$ is natural and epimorphic, and

$(\varepsilon^M_{XY'})': M(X,Y')' \to H(X,Y'')$ is easily checked up to coincide with

$\varphi^{M(.,..')'}_{XY}$, thus the latter is injective and $M(X,Y')'$ defines thus a bifunctor of type (II) and DM is therefore one of type (I).

4.12. Corollary: Any operator ideal of the form DM has the following property: Given $f \in H(X,Y)$, then $f \in DM(X,Y)$ iff $i_Y \circ f \in DM(X,Y'')$, and $\|f\|_{DM} = \|i_Y \circ f\|_{DM}$.

Proof: see 3.14.

4.13. Corollary: If M is a tensor product and is computable on the right hand side (i.e. $M(X,.)$ is computable for all X), then

$DM(X,Y') = M(X,Y)'$ for all X,Y'.

If M is a tensor product and Y' has the metric approximation property, then $DM(X,Y') = M(X,Y)'$

Proof: see 3.3, 3.4.

4.14. Given an operator ideal Λ, then we can consider the following norm on $X \otimes Y$:

$$\| \Sigma x_i \otimes y_i \|_{\Lambda^\otimes} = \sup \{ |\Sigma \langle y_i, f(x_i) \rangle |, \ f \in \Lambda (X,Y'), \ \|f\|_\Lambda \leq 1 \}.$$

It is a reasonable tensor norm, since if $\Lambda^\otimes(X,Y)$ denotes the completion, then we have

$$\Lambda^\otimes(X,Y) = (\Lambda(.,..')')_e(X,Y),$$

where $\Lambda(.,..')'_e$ is the partial functor of type Σ (4.4).

Proposition: If M is a tensor product,

$(DM)^\otimes (X,Y) = M(X,Y)$ holds if M is computable on the right-hand side, or if Y' has the metric approximation property.

Proof: Both conditions imply that

$DM(X,Y') = M(X,Y)'$. Then

$i_{M(X,Y)}: M(X,Y) \to M(X,Y)'' = (DM(X,Y'))'$ maps $M(X,Y)$, which is of type Σ, naturally and isometrically into the type Σ-part of $(DM(.,..'))'$ by an analogous result to 1.11, and has clearly dense image. qed.

Remark: When does $D(\Lambda^\otimes) = \Lambda$ hold? We will treat this question below.

4.15. One final result:

Theorem: Let Λ, Ω be operator ideals.

(a) $\underset{Ban^{op}}{Nat} (\Lambda(.,X), \Omega(.,Y)) =$

$\{f \in H(X,Y): f \circ g \in \Omega(Z,Y)$ for all $g \in \Lambda(Z,X)$, $Z \in Ban$ and

$\|f\|_{Nat} = \sup \{ \|f \circ g\|_\Omega, \ g \in \Lambda(Z,X), \ Z \in Ban, \ \|g\|_\Lambda \leq 1\} < \infty\}.$

and this defines again an operator ideal.

(b) $\underset{\text{Ban}}{\text{Nat}}\,(\Lambda(X,.),\,\Omega\,(Y,.))=$

$\{f \in H(X',Y'): f\circ g'$ *is weak*-continuous* $Z' \to Y'$, *its preadjoint*

$(f\circ g')^*$ *satisfies*

$(f\circ g')^* \in \Omega\,(Y,Z)$ *for all* $g \in \Lambda(X,Z)$, $Z \in$ Ban *and*

$$\|f\|_{\text{Nat}} = \sup\{\|f\circ g')^*\|_\Omega,\ g \in \Lambda(X,Z), Z \in \text{Ban},\ \|g\|_\Lambda \leq 1\} < \infty\}$$

and this defines a bifunctor of type (II).

Its associated operator ideal is given by

$\{f \in H(X,Y): g\circ f \in \Omega(Y,Z)$ *for all* $g \in \Lambda(X,Z), Z \in$ Ban *and*

$$\|f\|_{\text{Nat}} = \sup\{\|g\circ f\|_\Omega,\ g \in \Lambda(X,Z),\ Z \in \text{Ban},\ \|g\|_\Lambda \leq 1\} < \infty\}$$

Proof: (a): $\text{Nat}\,(\Lambda(.,I),\,\Omega(.,I)) = \text{Nat}\,(H(.,I),H(.,I))$

$= H(I,I) = I$ by Yoneda lemma.

The canonical map φ into $H(X,Y'')$ is given by $\varphi(\eta) = \eta_I$, $\eta \in \text{Nat}(\ldots)$, as checking up the definition shows. One proves injectivity of φ using the following computation:

For $z \in Z \in$ Ban and $f \in \Lambda\,(Z,X)$ we have

$$(\eta_Z(f))(z) = \Omega(\hat{z},Y)\,\eta_Z(f) = \eta_I \circ \Lambda(\hat{z},X)(f)$$

$$= \eta_I(f\circ\hat{z}) = \eta_I(f(z)).$$

Thus $\eta_Z(f) = \eta_I \circ f$, and $\eta_I = 0$ implies $\eta_Z = 0$ for all Z.

Thus we have got an operator ideal.

The relation $\eta_Z(f)=\eta_I \circ f$ shows that for all $\eta \in Nat(\ldots)$ the map η_I appears in the described set. Conversely, given f in that set, define a map $\Lambda(Z,X) \to \Omega(Z,Y)$ by $g \to f \circ g$ and the properties describing the set assure now that this map is a natural transformation, as a routine computation shows.

(b) is proven in a like manner; the canonical map φ is again given by $\eta \to \eta_I$ and the relevant computation for $\eta \in Nat(\ldots)$, $f \in \Lambda(X,Z)$, $z' \in Z'$ is the following

$$
\begin{aligned}
(\eta_Z(f))'(z') &= z' \circ (\eta_Z(f)) = \Omega(Y,z') \circ \eta_Z(f) \\
&= \eta_I \circ \Lambda(X,z')(f) = \eta_I(z' \circ f) \\
&= \eta_I \circ f'(z'),
\end{aligned}
$$

thus $\eta_Z(f)' = \eta_I \circ f'$; the rest is again routine matter, following the lines indicated above. qed.

§ 5. COMPUTABLE BIFUNCTORS AND MINIMAL OPERATOR IDEALS

5.1. Let G be a contra-covariant bifunctor with $G(I,I) = I$.
Then we define LG to be the bifunctor

$$LG(X,Y) = H(.,Y) \underset{(.)\in \mathrm{Fin}}{\hat{\otimes}} G(..,.) \underset{(..)\in \mathrm{Fin}}{\hat{\otimes}} H(X,..)$$

and we call G computable if $LG = G$ via the canonical mapping $LG \to G$. Thus G as a bifunctor is computable if all its partial functors $G(.,Y)$, $G(X,.)$ are computable (compare 4.7). Clearly L is left adjoint to the embedding of computable contra-covariant bifunctors, since we can check that up componentwise.
By 2.2 and 2.11 we have

$$LG(X,Y) = \lim \{G(X/M,E),\ X/M \in \mathrm{Fin},\ E \subseteq Y,\ E \in \mathrm{Fin}\}$$

(compare again 4.7).
LG is a contra-covariant bifunctor of type Σ (4.4) by 2.6 and 2.12.

5.2. Before beginning to treat minimal operator ideals we must deviate a little.
Counterexample 2.7 d) shows that although starting with an operator ideal Λ then $L\Lambda$ need not be again an operator ideal. But we can make it into one.
Let G be a contracovariant bifunctor with $G(I,I) = I$ and let $\varphi^G : G \to H(.,..'')$ be the canonical map (4.6, 4.8). Consider its canonical decomposition, componentwise:

$$\begin{array}{ccc} G(X,Y) & \xrightarrow{\varphi^{G}_{XY}} & H(X,Y'') \\ \text{coim } \varphi^{G}_{XY} \downarrow & & \uparrow \text{ im } \varphi^{G}_{XY} \\ \text{tot } G(X,Y) & \xrightarrow{\varphi^{G}_{XY}} & \overline{\varphi^{G}_{XY}(G(X,Y))}, \end{array}$$

(compare 1.12, where we used the same technique) where

$$\text{tot } G(X,Y) = G(X,Y)/(\varphi^{G}_{XY})^{-1}(0),$$

coim φ^{G}_{XY} is the quotient map,

$\overline{\varphi^{G}_{XY}(G(X,Y))}$ is the closure of the image of φ^{G}_{XY} in $H(X,Y'')$.

Since φ^{G}_{XY} is natural in X,Y, tot $G(X,Y)$ defines a contra-covariant bifunctor; since φ^{G} is natural in G the operation tot itself is a functor, which is furthermore left adjoint to the embedding of total bifunctors, i.e.

$\text{Nat}(G,G_1) = \text{Nat}(\text{tot } G, G_1)$ holds naturally in G and total G_1 (both satisfy of course $G(I,I) = I$, $G_1(I,I) = I$).

G is total if $G = \text{tot } G$ via coim φ^{G}.

5.3. Definition: Let Λ be an operatorideal.

We define $\Lambda^{\min}$ by

$$\Lambda^{\min} = \text{tot } (L\Lambda) ,$$

i.e. we restrict Λ to Fin, extend it by L and make it total.

$\Lambda^{\min}$ is then a total bifunctor and of type Σ since $L\Lambda$ is of type Σ. Thus $\Lambda^{\min}$ is an operator ideal.

We call Λ a minimal operator ideal, if $\Lambda = \Lambda^{min}$ via the canonical maps.

We will justify the name minimal by the next two results.

5.4. Proposition: (a) For all operator ideals we have

$$\Lambda^{min} = \Lambda^{min\ min}, \text{ i.e. } \Lambda^{min} \text{ is minimal.}$$

(b) tot $\circ$ L is left adjoint to the restriction of operator ideals on Fin.

Proof: (a) $\Lambda^{min\ min} = \text{tot} \circ L\,[(\text{tot} \circ L\,(\Lambda/\text{Fin}))|\text{Fin}]$

$= \text{tot} \circ L\,(\,L(\Lambda|\text{Fin})\,|\,\text{Fin}\,)$

since on Fin each functor is total,

$= \text{tot} \circ L(\Lambda|\text{Fin})$

$= \Lambda^{min}$

(b) Let Λ be an operator ideal on Fin and Ω be one on Ban. Then

$$\underset{\text{Fin} \times \text{Fin}}{\text{Nat}}\,(\Lambda, \Omega|\text{Fin}) =$$

$= \text{Nat}\,(L\Lambda, \Omega)$ by 5.1

$= \text{Nat}\,(\text{tot} \circ L(\Lambda), \Omega)$ by 5.2.

5.5. Theorem: An operator ideal Λ is minimal if and only if it is contractively contained in each operator ideal Ω with which it coincides on Fin.

The second condition in this theorem is equivalent to the definition of minimal Banach operator ideals of PIETSCH [19] 9.3.3, who, however, allows Ω to be any complete quasinormed operator ideal; for Λ a Banach operator ideal this is equivalent to our notion, since a quasinormed operator ideal of type Σ (i.e. operators of finite rank are dense) which is normed on finite-dimensional spaces is normed.

For the proof of this theorem we need a lemma.

Lemma: Let Λ, Ω be operator ideals, let $\underline{K}$ be a full regular (i.e. $X \in \underline{K}$ implies $X' \in \underline{K}$) subcategory of Ban. Then

$$\operatorname{Nat}_{\underline{K}^{op}\times\underline{K}} (\Lambda, \Omega) = I \text{ or } 0 .$$

It is I if the canonical map $\varphi^1 : \Lambda \to H$ takes its values in Ω and is bounded as a map $\Lambda \to \Omega$. Then $\operatorname{Nat}_{\underline{K}^{op}\times\underline{K}} (\Lambda, \Omega)$ is exactly the space of all scalar multiples of this canonical map.

Proof of the lemma: let be $0 \neq \psi \in \operatorname{Nat}_{\underline{K}\times\underline{K}} (\Lambda, \Omega)$.

Since $\underline{K}$ is regular, the map φ is defined.

We consider $\overset{\Omega}{\varphi} \circ \psi : \Lambda \to \Omega \to H$.

Take $f \in \Lambda(X,Y)$, then

$$\begin{aligned} < \varphi^{\Omega}_{XY} \circ \psi_{XY} (f)(x), y' > &= \Omega (\hat{x}, y') \psi_{XY} (f) \\ &= \psi_{II} \Lambda(\hat{x}, y') (f) \\ &= \psi_{II} <\varphi^{\Lambda} (f)(x), y'>. \end{aligned}$$

Since $\psi \neq 0$ there are X,Y such that $\psi_{XY} \neq 0$, then $\psi_{II} \neq 0$

since φ^{Ω} is injective. Thus $\psi_{II} = r \in H(I,I) = I$ and $\varphi^{\Omega} \circ \psi = r\cdot\varphi^{\Lambda}$, $r \neq 0$, thus φ^{Λ} takes its values in Ω and is a bounded map thereinto.

Proof of the theorem:

Let Ω be an operator ideal and let Λ be a minimal one with $\Lambda|\mathrm{Fin} = \Omega|\mathrm{Fin}$.

$$\text{then } \quad 1_{\Lambda|\mathrm{Fin}} \in \underset{\mathrm{Fin}^{op} \times \mathrm{Fin}}{\mathrm{Nat}} (\Lambda|\mathrm{Fin},\ \Omega|\mathrm{Fin})$$

$$= \underset{\mathrm{Ban}^{op} \times \mathrm{Ban}}{\mathrm{Nat}} (\mathrm{tot}\circ L(\Lambda|\mathrm{Fin}),\ \Omega) \quad \text{by } 5.4 \text{ (b)}$$

$$= \underset{\mathrm{Ban}^{op} \times \mathrm{Ban}}{\mathrm{Nat}} (\Lambda\ ,\ \Omega)\ .$$

Therefore the lemma implies that the canonical map $\varphi^{\Lambda}: \Lambda \to H$ factors through Ω and is an element of $\mathrm{Nat}\,(\Lambda,\Omega)$ with norm = 1 since it has the same norm as

$$1_{\Lambda|\mathrm{Fin}} \in \underset{\mathrm{Fin} \ \times \mathrm{Fin}}{\mathrm{Nat}} (\Lambda|\mathrm{Fin},\ \Omega|\mathrm{Fin})\ .$$

Conversely, let Λ be an operator ideal that is contractively contained in each Ω with $\Omega|\mathrm{Fin} = \Lambda|\mathrm{Fin}$. Thus Λ is contractively contained in $\Lambda^{\min}$, since $\Lambda|\mathrm{Fin} = \Lambda^{\min}|\mathrm{Fin}$, via the canonical map φ^{Λ}, but this map appears too as counit of the adjunction in 5.7 (b), it is therefore an isometric isomorphism.

<u>5.6.</u> <u>Example</u>: The bifunctor N_1 introduced in 2.13 d) which is actually an operator ideal, is minimal, since $\cdot' \hat{\otimes} \cdot\cdot$ is computable.

The operator ideal $\cdot' \hat{\hat{\otimes}} \cdot\cdot$, i.e. the norm closure of the maps of finite rank in H is minimal too.

We will give more examples later on.

§6. Complete functors and maximal operator ideals

6.1. We consider the category Ban^{Ban} of all covariant functors $F : Ban \to Ban$ and the restriction functor $F \to F|Fin$ onto Ban^{Fin}. This restriction functor $.|Fin$ has a left adjoint (see 2.2).

Proposition: The restriction functor $.|Fin$ for covariant functors has a right adjoint

$R : Ban^{Fin} \to Ban^{Ban}$; for $F : Fin \to Ban$

RF is given by

$$RF(X) = \operatorname{Nat}_{Fin} (H(X, .) , F) .$$

Proof: $F : Fin \to Ban$, $F_1 : Ban \to Ban$. Then we have

$$\operatorname{Nat}_{Ban} (F_1 , RF) =$$

$$= \operatorname{Nat}_{(.)\in Ban} (F_1(.) , \operatorname{Nat}_{(..)\in Fin} (H(.,..) , F(..)))$$

$$= \operatorname{Nat}_{(..)\in Fin} (H(.,..) \hat{\otimes}_{(.)\in Ban} F_1(.) , F(..)) \text{ by 1.8}$$

$$= \operatorname{Nat}_{(..)\in Fin} (F_1(..) , F(..)) \text{ by 1.13 a) since } Fin \subseteq Ban.$$

$$= \operatorname{Nat}_{Fin} (F_1|Fin, F) .$$

6.2. Proposition: For $X \in Ban$ and $F : Fin \to Ban$ we have

(a) $RF(X) = \lim \{F(X/M), X/M \in Fin\}$.

(b) $RF(X) = \operatorname{Nat}_{Ban} (X' \hat{\hat{\otimes}} . , F)$

(c) RF is always total, i.e. maps $RF(x')$, $x' \in X'$ separate points on $RF(X)$ for all X .

<u>Proof</u> : (a) similar to 2.11

(b) $RF(X) = \underset{Fin}{Nat}(X' \hat{\hat{\otimes}} ., F) = \underset{Ban}{Nat}(X' \hat{\hat{\otimes}} ., F)$

by 2.3 and 2.7 (b) .

(c) $\varphi_X^{RF}: RF(X) = \underset{Ban}{Nat}(X' \hat{\hat{\otimes}} ., F) \to H(X', F(I))$,

defined by $\varphi_X^{RF}(\xi)(x') = RF(x')(\xi)$ is easily seen to coincide with the map $\xi \to \xi_I$, $\xi \in RF(X)$, which is injective by the first part of the proof of theorem 3.2, since $X' \hat{\hat{\otimes}} .$ is of type Σ ; compare 4.6, 4.8, 5.2 for related discussions.

<u>6.3.</u> <u>Definition</u>: A functor $F : Ban \to Ban$ is said to be complete if $F = R(F|Fin)$ via the unit of the adjunction in 6.1, which is the map $\tau_X : F(X) \to \underset{Ban}{Nat}(X' \hat{\hat{\otimes}} ., F)$,

given by $(\tau_X^F(\xi))_Z (f) = F(f)\,\xi$ for $\xi \in F(X)$ and $f \in X' \hat{\hat{\otimes}} Z$.

If τ_X^F is isometric for all X , but not necessarily onto, then F is called a strong functor (see CIGLER [4]).

<u>6.4.</u> <u>Lemma</u>: <u>If</u> X <u>has the metric approximation property and</u> <u>if</u> F <u>is of type</u> Σ , <u>then</u> τ_X^F <u>is isometric</u>.

<u>Proof</u>: By 1.12 $K(X,X) = X' \hat{\hat{\otimes}} X$ has a left approximate identity (u_j) bounded by 1 and $F(X)$ is an essential left Banach - $K(X,X)$ - module. Thus for all $\xi \in F(X)$ we have

$$\| \tau_X^F (\xi)_X(u_j) - \xi \| = \| F(u_j)\, \xi - \xi \| \to 0 ,$$

thus $\| \tau_X^F (\xi) \| \geq \| \xi \|$; since clearly $\| \tau^F \| \leq 1$

we have $\| \tau_X^F (\xi) \| = \| \xi \|$.

6.5. Corollary: Let $F : \mathrm{Ban} \to \mathrm{Ban}$ be a functor. We write RF for $R(F|\mathrm{Fin})$ for short.

(a) $RF = RF_e$

(b) $R(DF)(X) = \underset{\mathrm{Ban}}{\mathrm{Nat}} (F, I_1(X', .))$

(c) $R(LF) = RF$

(d) $L(RF) = LF$

(e) $\underset{\mathrm{Ban}}{\mathrm{Nat}} (LF, F_1) = \underset{\mathrm{Ban}}{\mathrm{Nat}} (F, RF_1)$ naturally.

Proof: (a) $RF(X) = \underset{\mathrm{Ban}}{\mathrm{Nat}} (X' \hat{\hat{\otimes}} . \ , F)$ by 6.2 (b)

$= \underset{\mathrm{Ban}}{\mathrm{Nat}} (X' \hat{\hat{\otimes}} . \ , F_e)$ by 1.11

$= RF_e(X)$

(b) $R(DF)(X) = \underset{\mathrm{Ban}}{\mathrm{Nat}} (X' \hat{\hat{\otimes}} . \ , DF)$ by 6.2 (6)

$= \underset{\mathrm{Ban}}{\mathrm{Nat}} (F, D(X' \hat{\hat{\otimes}} .))$ by 3.1, Remark

$= \underset{\mathrm{Ban}}{\mathrm{Nat}} (F, I_1(X', .))$ by 3.2, Remark after the proof.

(c) $R(LF)(X) = \underset{\mathrm{Fin}}{\mathrm{Nat}} (X' \hat{\hat{\otimes}} ., (LF)|\mathrm{Fin})$

$= \underset{\mathrm{Fin}}{\mathrm{Nat}} (X' \hat{\hat{\otimes}} ., F|\mathrm{Fin}) = RF(X)$.

(d) $L(RF)(X) = (.'\hat{\hat{\otimes}} X) \underset{(.)\in Fin}{\hat{\otimes}} RF(.)$

$= (.'\hat{\hat{\otimes}} X) \underset{(.)\in Fin}{\hat{\otimes}} F(.) = LF(X)$,

since $RF|Fin = F|Fin$.

(e) $\underset{(.)\in Ban}{Nat} ((..'\hat{\hat{\otimes}} .) \underset{(..)\in Ban}{\hat{\otimes}} F(..), F_1(.)) =$

$= \underset{(..)\in Ban}{Nat} (F(..), \underset{(.)\in Ban}{Nat} ((..'\hat{\hat{\otimes}}.), F_1(.)))$ by 1.8

6.6. Examples:

(a) $R(X \hat{\otimes} .)(Y) = \underset{Ban}{Nat} (Y' \hat{\hat{\otimes}} ., X\hat{\otimes} .) =$

$= D(Y' \hat{\hat{\otimes}} .) (X) = I_1 (Y' , X)$,

compare 6.5, (b)

(b) $R(X \hat{\hat{\otimes}} .) (Y) = \underset{Ban}{Nat} (Y' \hat{\hat{\otimes}} ., X \hat{\hat{\otimes}} .)$

$= H(Y', X)$.

(c) $R(H(X,.))(Y) = R(H(X,.)_e)(Y)$ by 6.5 (a)

$= R(X' \hat{\hat{\otimes}} .) (Y)$

$= H(Y',X')$ by (b)

$= H(X,Y'')$.

(d) $R(I_1(X,.)) (Y) = R(I_1(X ,.)|Fin) (Y)$

$= R(X'\hat{\otimes} .|Fin)(Y)$, compare the proof of 3.3, where we used the same argument.

$= R(X' \hat{\otimes} .) (Y)$

$= I_1(Y',X') = (Y' \otimes X)'$

$= (X \hat{\hat{\otimes}} Y')' = I_1(X,Y'')$.

(e) A special case of any of the above results is: $R(Id) = ''$. By 6.2. (a) this amounts to $X'' = \lim \{X/M, X/M \in Fin\}$, which we already showed in 2.13(c).

Remark: One would not have expected $X \hat{\otimes}.$ to be complete, but (c), (d), (e) show that the completion of covariant functors behaves really bad; the completion of an operator ideal is thus of type (II), see 4.8.

6.7. We repeat the development for contravariant functors. We consider the restriction functor $.|Fin$ for contravariant functors $\overline{F}: Ban^{op} \to Ban$.

Proposition: The restriction functor $.|Fin$ for contravariant functors has a right adjoint

$R: Ban^{Fin^{op}} \to Ban^{Ban^{op}}$; for $\overline{F}: Fin^{op} \to Ban$

$R\overline{F}$ is given by

$$R\overline{F}(X) = \underset{Fin}{Nat}\,(H(.,X),\overline{F}).$$

Proof: like 6.1.

6.8. Proposition: For $X \in Ban$ and $\overline{F}: Fin^{op} \to Ban$ we have:

(a) $R\overline{F}(X) = \lim \{\overline{F}(E), E \subset X, E \in Fin\}$.

(b) $R\overline{F}(X) = \underset{Ban}{Nat}\,(.' \hat{\hat{\otimes}} X, \overline{F})$.

(c) $R\overline{F}$ is always total, i.e. maps $R\overline{F}(\hat{x})$, $x \in X$ separate points on $R\overline{F}(X)$ for all X.

Proof: like 6.2.

6.9. Definition A functor $\overline{F}$: $Ban^{op} \to Ban$ is said to be complete, if $\overline{F} = R(\overline{F}|Fin)$ via the unit of the adjunction in 6.5. i.e. the map $\tau_X^{\overline{F}}: \overline{F}(x) \to \underset{Ban}{Nat}(.'\hat{\hat{\otimes}} X, \overline{F})$, given by

$$(\tau_X^{\overline{F}}(\xi))_Z(f) = \overline{F}(f)\,\xi \text{ for } \xi \in \overline{F}(X),\ f \in Z'\hat{\hat{\otimes}} X.$$

If $\tau_X^{\overline{F}}$ is isometric only for all X, then we call $\overline{F}$ a strong functor. (See CIGLER [4]).

RF, $R\overline{F}$ is sometimes called the completion of the functor F, $\overline{F}$.

6.10 Corollary: Let be $\overline{F}$: $Ban^{op} \to Ban$.

(a) $R\overline{F} = R\overline{F}_e$

(b) $R(L\overline{F}) = R\overline{F}$

(c) $L(R\overline{F}) = L\overline{F}$

(d) $\underset{Ban}{Nat}(L\overline{F}, \overline{F}_1) = \underset{Ban}{Nat}(\overline{F}, R\overline{F}_1)$ naturally.

Proof: like 6.5.

6.11 Examples:

(a) $R(X\hat{\otimes}.')(Y) = \underset{Ban}{Nat}(.'\hat{\hat{\otimes}} Y, .'\hat{\otimes} X)$

$= I_1(Y,X)$ by using 3.6.

(b) $R(X\hat{\hat{\otimes}}.')(Y) = \underset{Ban}{Nat}(.'\hat{\hat{\otimes}} Y, .'\hat{\hat{\otimes}} X)$

$= H(Y,X)$

(c) $R(H(.,X)) = R(H(.,X)_e) = R(.'\hat{\hat{\otimes}} X) = H(.,X)$.

(d) $R(.') = .'$ by e.g. (c).

Remark: So contravariant functors have rather well behaved completions. Remind that the contravariant part offered no complication at all in defining operator ideals.

We will add considerably to the examples of covariant and contravariant complete functors in §7.

6.12 Let G be a bifunctor $\mathrm{Fin}^{op} \times \mathrm{Fin} \to \mathrm{Ban}$ satisfying $G(I,I) = I$. We write RG for the functor: $\mathrm{Ban}^{op} \times \mathrm{Ban} \to$ given by

$$RG(X,Y) = \underset{(.)\in \mathrm{Fin}}{\mathrm{Nat}} (Y' \hat{\hat{\otimes}} ., \underset{(..)\in \mathrm{Fin}}{\mathrm{Nat}} (..' \hat{\hat{\otimes}} X, G(..,.)))$$

$$= \underset{(.),(..)\in \mathrm{Fin} \times \mathrm{Fin}}{\mathrm{Nat}} ((Y' \hat{\hat{\otimes}} .) \hat{\otimes} (..' \hat{\hat{\otimes}} X), G(..,.))$$

$$= \underset{(..)\in \mathrm{Fin}}{\mathrm{Nat}} (..' \hat{\hat{\otimes}} X, \underset{(.)\in \mathrm{Fin}}{\mathrm{Nat}} (Y' \hat{\hat{\otimes}} ., G(..,.))).$$

The little computation we just did shows that

$RG(X,Y) = R^{(.)} R^{(..)} G(..,.) = R^{(..)} R^{(.)} G(..,.)$ holds or that we have:

$$RG(X,Y) = \lim \{\lim \{G(E,Y/M),\ Y/M \in \mathrm{Fin}\},\ E \subset X,\ E \in \mathrm{Fin}\}$$

$$= \lim \{\lim \{G(E,Y/M),\ E \subset X,\ E \in \mathrm{Fin}\},\ Y/M \in \mathrm{Fin}\}$$

$$= \lim \{G/E,Y/M),\ E \subset X,\ E \in \mathrm{Fin},\ Y/M \in \mathrm{Fin}\}.$$

RG is always total by 6.2(c) and 6.6(c).

Proposition: R is right adjoint to the restriction of bifunctors G with $G(I,I) = I$ to Fin, i.e.

$$\underset{\mathrm{Fin}^{op} \times \mathrm{Fin}}{\mathrm{Nat}} (G|\mathrm{Fin}, G_1) = \underset{\mathrm{Ban}^{op} \times \mathrm{Ban}}{\mathrm{Nat}} (G, RG_1)$$ holds naturally in

G and G_1.

Proof: Combine 6.1 and 6.7.

6.13. Definition: Let Λ be an operator ideal. We define Λ^{max} by $\Lambda^{max} = (R\Lambda)^{(I)}$ (recall 4.10), i.e. we restrict Λ to Fin, extend it again and make it an operator ideal.

Clearly Λ^{max} is then an operator ideal, since $R\Lambda$ is total and $R\Lambda(I,I): \Lambda(I,I) = I$, thus $(R\Lambda)^{(I)}$ is of type (I).

We call Λ a maximal operator ideal, if $\Lambda = \Lambda^{max}$ via the canonical maps.

The name maximal will be justified by the next two results.

6.14. Proposition: (a) $\Lambda^{max\ max} = \Lambda^{max}$, i.e. Λ^{max} is maximal for all operator ideals Λ.

(b) $(R.)^{(I)}$ is right adjoint to the restriction of operator ideals to Fin.

Proof: (a) $R(\Lambda|Fin)^{(I)}|Fin = R(\Lambda|Fin)Fin$, since all functors with $G(I,I) = I$ are of type (I) an Fin, then $R(\Lambda|Fin)|Fin = \Lambda|Fin$, then $\Lambda^{max\ max} = R(R(\Lambda|Fin)^{(I)}|Fin)^{(I)} = R(\Lambda|Fin)^{(I)} = \Lambda^{max}$.

(b) Yet Ω be an operator ideal and Λ be a bifunctor of type (I) on Fin. Then

$$\mathrm{Nat}_{Fin^{op} \times Fin}(\Omega|Fin,\Lambda) = \mathrm{Nat}_{Ban^{op} \times Ban}(\Omega,R\Lambda) \text{ by 6.10}$$

$$= \mathrm{Nat}_{Ban^{op} \times Ban}(\Omega,(R\Lambda)^{(I)}) \text{ by 4.10.}$$

qed

6.15. Theorem: An operator ideal Λ is maximal if and only if each operator ideal Ω that coincides with Λ on Fin is contained contractively in Λ.

The second condition of this theorem is equivalent to the definition of PIETSCH [19] 9.3.3 for maximal Banach operator ideals, as is easily seen, taking convex hulls of unit balls.

Proof: If Λ is maximal, i.e. $\Lambda = \Lambda^{max}$, and if Ω is an operator ideal with $\Omega|Fin = \Lambda|Fin$, then $1_{\Lambda|Fin} \in \underset{Fin^{op}\times Fin}{Nat}(\Omega|Fin, \Lambda|Fin)$:

$$= \underset{Ban^{op}\times Ban}{Nat}(\Omega, \Lambda^{max}) = \underset{Ban^{op}\times Ban}{Nat}(\Omega, \Lambda).$$

Thus by the lemma in 5.5 the canonical map $\varphi^{\Omega} : \Omega \to H$ takes its values in Λ and is bounded by 1 as a map $\Omega \quad \Lambda$ since it has the same norm as its corresponding element $1_{\Lambda|Fin} \in \underset{Fin^{op}\times Fin}{Nat}(\Omega|Fin, \Lambda|Fin)$, but this means Ω is contractively contained in Λ.

Conversely, let us suppose that Λ contains contractively each operator ideal Ω with $\Lambda|Fin = \Omega|Fin$. But then $\Lambda^{max}|Fin = \Lambda|Fin$, thus Λ^{max} is contractively contained in Λ; Λ itself however is contractively contained in Λ^{max} via the unit of the adjunction of 6.12 (b), thus $\Lambda = \Lambda^{max}$. qed

6.16. Examples:

(a) H is maximal: $(RH)^{(I)} = H(.,..'')^{(I)} = H$.

(b) I_1 is maximal $(RI_1)^{(I)} = I_1(.,..'')^{(I)} = I_1$.

We will produce more examples later on.

6.18. Lemma: Let F: Ban $\to$ Ban and $\overline{F}$: $\mathrm{Ban}^{\mathrm{op}} \to$ Ban be functors: F is computable if and only if F' is complete. $\overline{F}$ is computable if and only if $\overline{F}'$ is complete.

Proof: The unit of the adjunction 6.1:

$$\tau_X^{F'}: F(X)' \to \underset{\mathrm{Fin}}{\mathrm{Nat}}\,(.' \hat{\hat{\otimes}} X, F') = ((.' \hat{\hat{\otimes}} X) \underset{\mathrm{Fin}}{\hat{\otimes}} F)'$$

is easily seen to be the adjoint of the counit of the adjunction 2.9:

$$(.' \hat{\hat{\otimes}} X) \underset{\mathrm{Fin}}{\hat{\otimes}} F \to F(X).$$

Thus the first of these maps is an isometric if and only if the second one is it.

The method works for the contravariant case too.

6.18. Corollary: Let M be a tensor product (4.7). M is computable if and only if the operator ideal DM (4.11) is maximal

Proof. 6.15 for necessity. If DM is maximal, then DM(.,..') = M' since DM(.,..')|Fin = M'|Fin, Thus M' is complete (check this) and again by 6.15 M is computable. qed.

§7. The projective (p,r,s)-tensor product

7.1. First of all we introduce some norms for sequences in Banach spaces: let X be a Banach space and let $(x_i)_{i=1}^{\infty}$ be a sequence in X (finite sequences are thought to be continued by zeros). Then we will consider the following norms:

$$\| (x_i) \|_{\ell^p} \ (\sum_i \| x_i \|^p)^{1/p}, \quad 0 < p < \infty$$

$$\| (x_i) \|_{\ell^\infty} = \sup_i \| x_i \|$$

$$\| (x_i) \|_{\varepsilon^p} = \sup_{\| x' \| \leq 1} (\sum_i | \langle x_i, x' \rangle |^p)^{1/p}, \quad 0 < p < \infty$$

$$\| (x_i) \|_{\varepsilon^\infty} = \sup_{\| x' \| \leq 1} (\sup_i | \langle x_i, x' \rangle |).$$

It is immediate to check that $\| \cdot \|_{\varepsilon^\infty} = \| \cdot \|_{\ell^\infty}$ holds.
For $1 \leq p \leq \infty$ we can consider the space $\varepsilon^p(X)$ of all sequences $(x_i) \subset X$ which satisfy $\| (x_i) \|_{\varepsilon^p} < \infty$. This space turns out to be a Banach space and with coordinate wise action we get a functor ε^p. Ban $\to$ Ban. It is a routine matter to verify that $\varepsilon^p(X) = H(\ell^{p'}, X)$ holds, where $1/p + 1/p' = 1$. Thus clearly $\varepsilon^p(f)$ is isometric whenever f is and this property holds too in case $0 < p < 1$, where it can be verified by direct computation.

7.2 Theorem:

Let X, Y be Banach spaces and $0 < p,r,s, \leq \infty$ such that $1 \leq 1/q = 1/p + 1/r + 1/s < \infty$. For $u \in X \otimes Y$ define

$|||u|||_{(p,r,s)} = \inf \|(\lambda_i)\|_{\ell^p} \|(x_i)\|_{\varepsilon^r} \|(y_i)\|_{\varepsilon^s}$, where

the infimum is taken over all representations

$$u = \sum_i \lambda_i x_i \otimes y_i \text{ in } X \otimes Y.$$

If $q = 1$, then the expression $|||\,.\,|||_{(p,r,s)}$ is a bifunctorial reasonable crossnorm on $X \otimes Y$, which we design by $\|\,.\,\|_{(p,r,s)}$. The completion $X \hat{\otimes}_{(p,r,s)} Y$ in this norm gives a tensor product, which turns out to be computable.

If $0 < q < 1$, then the expression $|||\,.\,|||_{(p,r,s)}$ is a q-norm on $X \otimes Y$ (i.e. satisfies the condition of 7.3. instead of the triangle inequality). The Minkovski functional $\|\,.\,\|_{(p,r,s)}$ of the convex hull of the "unit ball" $\{u \in X \otimes Y: |||u|||_{(p,r,s,} \leq 1\}$ however is a bifunctorial, reasonable norm on $X \otimes Y$; the completion $X \hat{\otimes}_{(p,r,s)} Y$ in this norm gives a tensor product, which turns out to be computable.

We call $\hat{\otimes}_{(p,r,s)}$ the projective (p,r,s)-tensor product.

7.3. Lemma: For $u_1, u_2 \in X \otimes Y$ we have

$$(|||u_1 + u_2|||_{(p,r,s)})^q \leq (|||u_1|||_{(p,r,s)})^q + (|||u_2|||_{(p,r,s)})^q$$

Proof: Let be $\varepsilon > 0$ and let $u_j = \sum_i \lambda_{ij} x_{ij} \otimes y_{ij}$, $j = 1,2$, be representations such that the following holds

$$(\|(\lambda_{ij})_i\|_{\ell^p}\ \|(x_{ij})_i\|_{\varepsilon^r}\ \|(y_{ij})_i\|_{\varepsilon^s})^q \leqslant (|||u_j|||_{(p,r,s)})^q + \varepsilon, j=1,2.$$

By shifting scalars we can suppose that

$$\|(\lambda_{ij})_i\|_{\ell^p}^q \leqslant (|||u_j|||_{(p,r,s)}^q + \varepsilon)^{q/p}$$

$$\|(x_{ij})_i\|_{\varepsilon^r}^q \leqslant (|||u_j|||_{(p,r,s)}^q + \varepsilon)^{q/r}$$

$$\|(y_{ij})_i\|_{\varepsilon^s}^q \leqslant (|||u_j|||_{(p,r,s)}^q + \varepsilon)^{q/s}.$$

Bear in mind that $0 < q \leqslant 1$. Cases $p, r, s = \infty$ simply mean that the right-hand side is 1, since we may suppose $u_j \neq 0$, $j = 1,2$. Then:

$$u_1 + u_2 = \sum_{j=1}^{2} \sum_i \lambda_{ij}\, x_{ij} \otimes y_{ij},$$

and for $p, r < \infty$ we have:

$$\|(\lambda_{ij})\|_{\ell^p}^q = (\sum_i |\lambda_{i1}|^p + \sum_i |\lambda_{i2}|^p)^{q/p}$$

$$= (\|(\lambda_{i1})_i\|_{\ell^p}^p + \|(\lambda_{i2})_i\|_{\ell^p}^p)^{q/p}$$

$$\leqslant (|||u_1|||_{(p,r,s)}^q + \varepsilon + |||u_2|||_{(p,r,s)}^q + \varepsilon)^{q/p},$$

$$\|(x_{ij})\|_{\varepsilon^r}^q = \sup_{\|x'\|\leqslant 1} (\sum_i |\langle x_{i1}, x'\rangle|^r + \sum_i |\langle x_{i2}, x'\rangle|^r)^{q/r}$$

$$\leqslant (\|(x_{i1})_i\|_{\varepsilon^r}^r + \|(x_{i2})_i\|_{\varepsilon^r}^r)^{q/r}$$

$$\leqslant (|||u_1|||_{(p,r,s)}^q + \varepsilon + |||u_2|||_{(p,r,s)}^q + \varepsilon)^{q/r},$$

and the same estimate trivially holds for $p, r, = \infty$. Thus we can compute:

$$|||u_1 + u_2|||_{(p,r,s)}^q \leq \|(\lambda_{ij})\|_{\ell^p}^q \|(x_{ij})\|_{\varepsilon^r}^q \|(y_{ij})\|_{\varepsilon^s}^q$$

$$\leq (|||u_1|||_{(p,r,s)}^q + |||u_2|||_{(p,r,s)}^q + 2\varepsilon)^{q/p+q/r+q/s}$$

$$= |||u_1|||_{(p,r,s)}^q + |||u_2|||_{(p,r,s)}^q + 2\varepsilon \qquad \text{qed}$$

__7.4.__ __Lemma:__ $|||\cdot|||_{(p,r,s)}$ __is positive homogenous and__

$\|u\|^\wedge \leq |||u|||_{(p,r,s)}$ __for all__ $u \in X \otimes Y$.

__Proof:__ The first assertion is immediate.

For the second let $u = \sum_i \lambda_i x_i \otimes y_i$ be a representation in $X \otimes Y$. Then:

$$\|u\|^\wedge = \sup_{\|x'\| \leq 1, \|y'\| \leq 1} |\sum_i \lambda_i \langle x_i, x'\rangle \langle y_i, y'\rangle|$$

$$\leq \sup_{\|x'\| \leq 1, \|y'\| \leq 1} \|(\lambda_i)\|_{\ell^{p/q}} \cdot \|(\langle x_i, x'\rangle)\|_{\ell^{r/q}} \cdot \|(\langle y_i, y'\rangle)\|_{\ell^{s/q}},$$

by the Hölder inequality, since $q/p + q/r + q/s = 1$

$$\leq \|(\lambda_i)\|_{\ell^p} \cdot \sup_{\|x'\| \leq 1} \|(\langle x_i, x'\rangle)\|_{\ell^r} \cdot \sup_{\|y'\| \leq 1} \|(\langle y_i, y'\rangle)\|_{\ell^s},$$

since $p \leq p/q$, thus $\|(\lambda_i)\|_{\ell^{p/q}} \leq \|(\lambda_i)\|_{\ell^p}$ etc,

because by multiplying with a scalar we can assume that $\sum_i |\lambda_i|^p = 1$, then $|\lambda_i|^p \leq 1$ for all i, $|\lambda_i|^{p/q} \leq |\lambda_i|^p$, since $1/q \geq 1$, thus $\sum_i |\lambda_i|^{p/q} \leq 1$ and $\|(\lambda_i)\|_{\ell^{p/q}} \leq 1 = \|(\lambda_i)\|_{\ell^p}$

Now

$\|u\|^\wedge \leq \|(\lambda_i)\|_{\ell^p} \|(x_i)\|_{\varepsilon^r} \|(y_i)\|_{\varepsilon^s}$ and therefore

$\|u\|^\wedge \leq |||u|||_{(p,r,s)}$ qed

<u>7.5. Lemma</u>: $\| \cdot \|_{(p,r,s)}$ <u>(as defined in 7.2) is a bifunctorial reasonable crossnorm on</u> $X \otimes Y$

<u>Proof</u>: First we consider the case $q = 1$.

$\| \cdot \|_{(p,r,s)} = ||| \cdot |||_{(p,r,s)}$ satisfies the triangle inequality by 7.3, thus for any representation $u = \sum_i x_i \otimes y_i$ in $X \otimes Y$ we have

$$\| u \|_{(p,r,s)} \leq \sum_i \| x_i \otimes y_i \|_{(p,r,s)}$$
$$\leq \sum_i \| x_i \| \, \| y_i \| ,$$

therefore $\| u \|_{(p,r,s)} \leq \| u \|^{\wedge}$ (see 1.2).

If $\| u \|_{(p,r,s)} = 0$ then $\| u \|^{\wedge} = 0$ too by 7.4 and $u = 0$; so $\| \cdot \|_{(p,r,s)}$ is a norm and satisfies $\| \cdot \|^{\wedge} \leq \| \cdot \|_{(p,r,s)} \leq \| \cdot \|^{\wedge}$

i.e. it is a reasonable norm (1.9).

Given $f \in H(X,X_1)$, $g \in H(Y,Y_1)$ and a representation $u = \sum_i \lambda_i x_i \otimes y_i$ in $X \otimes Y$, then

$$\| (f \otimes g)u \|_{(p,r,s)} = \| \sum_i \lambda_i \, f(x_i) \otimes g(y_i) \|_{(p,r,s)}$$
$$\leq \|(\lambda_i)\|_{\ell^p} \, \|(fx_i)\|_{\varepsilon^r} \, \|(gy_i)\|_{\varepsilon^s}$$
$$\leq \|f\| \, \|g\| \, \|(\lambda_i)\|_{\ell^p} \, \|(x_i)\|_{\varepsilon^r} \, \|(y_i)\|_{\varepsilon^s} \text{ by 7.1.}$$

So $\|.\|_{(p,r,s)}$ is bifunctorial too.

Now we treat the case $0 < q < 1$.

Write $M = \{u \in X \otimes Y: |||u|||_{(p,r,s)} \leqslant 1\}$.

Since $\|.\|^{\hat{\hat{}}} \leqslant |||.|||_{(p,r,s)}$ we have $M \subset O(X \hat{\hat{\otimes}} Y) \cap (X \otimes Y)$.

Thus the absolutely convex hull ΓM too satisfies $\Gamma M \subset O(X \hat{\hat{\otimes}} Y) \cap (X \otimes Y)$, since the latter set is convex. It is well known (see e.g. GROTHENDIECK [7]) that $O(X \hat{\otimes} Y) \cap (X \otimes Y)$ is the convex hull of the set $P = \{x \otimes y, \|x\| \leqslant 1, \|y\| \leqslant 1\}$ in $X \otimes Y$. Since $P \subset M$ we have $O(X \hat{\otimes} Y) \cap (X \otimes Y) = \Gamma P \subset \Gamma M \subset O(X \hat{\hat{\otimes}} Y) \cap (X \otimes Y)$.

Now let $\|.\|_{(p,r,s)}$ be the Minkovski functional of ΓM, i.e.

$$\|u\|_{(p,r,s)} = \inf \{r > 0, \tfrac{1}{r}.u \in \Gamma M\}.$$

The above chain of inclusion then implies that $\|.\|^{\hat{\hat{}}} \leqslant \|.\|_{(p,r,s)} \leqslant \|.\|^{\wedge}$; $\|.\|_{(p,r,s)}$ is a norm since ΓM is absolutely convex and by the above inclusions. Thus $\|.\|_{(p,r,s)}$ is a reasonable norm on $X \otimes Y$.

Now let be $f \in H(X,X_1)$, $g \in H(Y,Y_1)$; then

$|||(f \otimes g)\, u|||_{(p,r,s)} \leqslant \|f\|\, \|g\|\, |||u|||_{(p,r,s)}$, $u \in X \otimes Y$, by the above computation. That means

$(f \otimes g)(M_{X \otimes Y}) \subset \|f\| . \|g\| . M_{X_1 \otimes Y_1}$, so

$(f \otimes g)\, (\Gamma M_{X \otimes Y}) = \Gamma(f \otimes g)(M_{X \otimes Y}) \subset \|f\| . \|g\| . \Gamma M_{X_1 \otimes Y_1}$,

i.e. $\|(f \otimes g)u\|_{(p,r,s)} \leqslant \|f\| \|g\| \|u\|_{(p,r,s)}$ for $u \in X \otimes Y$ and so $\|.\|_{(p,r,s)}$ is bifunctorial too. qed.

<u>7.6. Lemma: The tensor product $\cdot\hat{\otimes}_{(p,r,s)}\cdot$ is computable.</u>

<u>Proof</u>: Since clearly $X \hat{\otimes}_{(p,r,s)} Y = Y \hat{\otimes}_{(p,s,r)} X$ it suffices to show that the functor $X \hat{\otimes}_{(p,r,s)}\cdot$ is computable for all X. We use 2.2.

We consider the spectral family $\{X \hat{\otimes}_{(p,r,s)} E, E \subset Y, E \in \text{Fin}\}$

Let $\{f_E : X \hat{\otimes}_{(p,r,s)} E \to Z, E \subset Y, E \in \text{Fin}\}$ be a map from this spectral family into an arbitrary Banach space Z, i.e.

$\|f_E\| \leq 1$ for all E and $f_E = f_{E_1} \circ (X \hat{\otimes}_{(p,r,s)} i_E^{E_1})$, where $i_E^{E_1}$ is the embedding $E \to E_1$.

We have to find a uniquely determined map $f: X \hat{\otimes}_{(p,r,s)} Y \to Z$ with $\|f\| \leq 1$ and $f_E = f \circ (X \hat{\otimes}_{(p,r,s)} i_E)$ where $i_E : E \to Y$ is the embedding.

Given $u \in X \otimes Y$ and any representation $u = \sum_{i=1}^{n} \lambda_i x_i \otimes y_i$, take a finite-dimensional subspace $E \subset Y$ such that all $y_i \in E$. Then we should have

$$f(u) = f_E (\Sigma \lambda_i x_i \otimes y_i).$$

If we define f in that way we should note first that the definition is independent of the choice of $E \subset Y$: if $\{y_i\} \subset E, \{y_i\} \subset E_1$, put $E_2 = E + E_1$, then

$$\begin{aligned} f_E (\Sigma \lambda_i x_i \otimes y_i) &= f_{E_2} \circ (X \hat{\otimes}_{(p,r,s)} i_E^{E_2}) (\textstyle\sum_i \lambda_i x_i \otimes y_i) \\ &= f_{E_2} \circ (X \otimes_{(p,r,s)} i_{E_1}^{E_2})(\textstyle\sum_i \lambda_i x_i \otimes y_i) \\ &= f_{E_1} (\Sigma \lambda_i x_i \otimes y_i). \end{aligned}$$

A similar argument shows that $f(u)$ is independent of the representation of u too, thus $f: X \otimes Y \to Z$ is linear and uniquely determined and $f|X \otimes E = f_E$.

It remains to show that $\|f\| \leq 1$.

We consider first the case $q = 1$.

Let $u = \sum_i \lambda_i x_i \otimes y_i$ be any representation in $X \otimes Y$. Take $\{y_i\} \subset E \subset Y$. Then:

$$\|f(u)\|_Z = \|f_E(u)\|_Z \leq \|f_E\| \, \|u\|_{X \hat{\otimes}_{(p,r,s)} E}$$

$$\leq 1. \, \|(\lambda_i)\|_{\ell^p} \, \|(x_i)\|_{\varepsilon^r(X)} \|(y_i)\|_{\varepsilon^s(E)}.$$

By 7.1. we know that $\varepsilon^s(i_E)$ is isometric, so

$\|(y_i)\|_{\varepsilon^s(E)} = \|(y_i)\|_{\varepsilon^s(Y)}$ and

$\|f(u)\|_Z \leq \|(\lambda_i)\|_{\ell^p} \, \|(x_i)\|_{\varepsilon^r(X)} \|(y_i)\|_{\varepsilon^s(Y)}$ holds for any representation, i.e. $\|f(u)\|_Z \leq \|u\|_{(p,r,s)}$ or $\|f\| \leq 1$.

Now we consider the general case. We had $O(X \hat{\otimes}_{(p,r,s)} Y) \cap (X \otimes Y) = \Gamma M$, where $M = \{u \in X \otimes Y: |||u|||_{(p,r,s)} \leq 1\}$, cf. 7.5.

$$\|f\| = \sup_{u \in O(X \hat{\otimes}_{(p,r,s)} Y)} \|f(u)\|$$

$$= \sup_{u \in \Gamma M} \|f(u)\|$$

$$= \sup_{u \in M} \|f(u)\| \text{ as is easily seen.}$$

Let be $u \in M$ and $\varepsilon > 0$. Take a representation $u = \Sigma\, \lambda_i\, x_i \otimes y_i$ such that

$$||(\lambda_i)||_{\ell^p}\ ||(x_i)||_{\varepsilon^r(X)} ||(y_i)||_{\varepsilon^s(Y)} \leq |||u|||_{(p,r,s)} + \varepsilon \leq 1 + \varepsilon$$

Take any $E \subset Y$, $E \in$ **Fin** with $\{y_i\} \subset E$. By 7.1 we know that $\varepsilon^s(i_E)$ is isometric, so

$$||(y_i)||_{\varepsilon^s(Y)} = ||(y_i)||_{\varepsilon^s(E)}.$$

Consequently $||(\lambda_i)||_{\ell^p}\ ||(x_i)||_{\varepsilon^r(X)}\ ||(y_i)||_{\varepsilon^s(E)} \leq 1 + \varepsilon$,

i.e. $u = \Sigma\, \lambda_i x_i \otimes y_i \in (1 + \varepsilon)\, M_{X \otimes E} \subset (1 + \varepsilon) O(X \hat{\otimes}_{(p,r,s)} E)$

and therefore we have $||f(u)|| = ||f_E(u)|| \leq ||f_E||\ ||u||_{X \hat{\otimes}_{(p,r,s)} E} \leq 1+\varepsilon$

for any $u \in M$ and $\varepsilon > 0$, so $||f'|| \leq 1$. qed.

Theorem 7.2. is now completely proved.

<u>7.7.</u> <u>Lemma</u>: $X \hat{\otimes}_{(p,r,s)} \cdot'$ <u>is computable for all</u> X.

<u>Proof</u>: $Y' = \underset{\rightarrow}{\lim}\ \{(Y/M)',\ Y/M \in \mathbf{Fin}\}$

$\quad = \underset{\rightarrow}{\lim}\ \{E,\ E \subset Y',\ E \in \mathbf{Fin}\}$ by 2.13 (a), (b).

$X \hat{\otimes}_{(p,r,s)} Y' = X \hat{\otimes}_{(p,r,s)} (\underset{\rightarrow}{\lim}\ \{E,\ E \subset Y',\ E \in \mathbf{Fin}\})$

$\quad = \underset{\rightarrow}{\lim}\ \{X \hat{\otimes}_{(p,r,s)} E,\ E \subset Y',\ E \in \mathbf{Fin}\}$ by 7.6, 2.2,

$\quad = \underset{\rightarrow}{\lim}\ \{X \hat{\otimes}_{(p,r,s)} (Y/M)',\ Y/M \in \mathbf{Fin}\}$

thus $X \hat{\otimes}_{(p,r,s)} \cdot'$ is computable by 2.11. qed.

7.8. Remarks:

(a) If $0 < p, r, s \leq \infty$ and $1/q = 1/p + 1/r + 1/s < 1$, then we have $||.||_{(p,r,s)} = 0$, since for $x \otimes y \in X \otimes Y$ we have

$$||x \otimes y||_{(p,r,s)} = ||\sum_{i=1}^{n} 1/n \cdot x \otimes y||_{(p,r,s)}$$

$$\leq ||(1/n)_{i=1}^{n}||_{\ell^p} \; ||(x)_{i=1}^{n}||_{\varepsilon^r} \; ||(y)_{i=1}^{n}||_{\varepsilon^s}$$

$$= n^{1/p+1/r+1/s-1} \cdot ||x|| \; ||y|| \to 0 \text{ for } n \to \infty.$$

(b) Define

$$|||u|||_{(\ell^p,\ell^r,\varepsilon^s)} = \inf ||(\lambda_i)||_{\ell^p} \; ||(x_i)||_{\ell^r} \; ||(y_i)||_{\varepsilon^s},$$

$$|||u|||_{(\ell^p,\varepsilon^r,\ell^s)} = \inf ||(\lambda_i)||_{\ell^p} \; ||(x_i)||_{\varepsilon^r} \; ||(y_i)||_{\ell^s},$$

$$|||u|||_{(\ell^p,\ell^r,\ell^s)} = \inf ||(\lambda_i)||_{\ell^p} \; ||(x_i)||_{\ell^r} \; ||(y_i)||_{\ell^s},$$

where the infimum is always taken over all representations $u = \Sigma \lambda_i x_i \otimes y_i$ in $X \otimes Y$. Then we have

$$|||.|||_{(\ell^p,\ell^r,\varepsilon^s)} = |||.|||_{(1/(1/p + 1/r), \infty, s)},$$

$$|||.|||_{(\ell^p,\varepsilon^r,\ell^s)} = |||.|||_{(1/(1/p + 1/s), r, \infty)},$$

$$|||.|||_{(\ell^p,\ell^r,\ell^s)} = |||.|||_{(q,\infty,\infty)}, \; 1/q = 1/p + 1/r + 1/s.$$

(c) Using (b) and the fact that $||.||_{(1,\infty,\infty)} = ||.||^{\wedge}$ it is immediately clear that for $1 \leq p \leq \infty$, $1/p + 1/p' = 1$ the norm $||.||_{(p,\infty,p')}$ coincides with the norm g_p of SAPHAR, [20], §3 and with the p-norm of CHEVET [2]. The norm $||.||_{(p,p',\infty)}$ coincides with the norm d_p of SAPHAR and the ${}^t p$-norm of CHEVET. An immediate consequence is:

<u>Theorem: Let $(\Omega_1, \Sigma_1, \mu_1)$ and $(\Omega_2, \Sigma_2, \mu_2)$ be two σ-finite measure spaces. Then we have</u> $L^p(\Omega_1) \hat{\otimes}_{(p,p',\infty)} L^p(\Omega_2)$

$$= L^p(\Omega_1 \times \Omega_2, \mu_1 \times \mu_2)$$

$$= L^p(\Omega_1, L^p(\Omega_2)) = L^p(\Omega_2, L^p(\Omega_1)).$$

See CHEVET [2].

<u>Proof</u>: of (b): The first assertion and the second one are essentially the same, we prove the first one.

Suppose $p, r \neq \infty$, $u \in X \otimes Y$ and let $u = \sum_i \lambda_i x_i \otimes y_i$ be a representation. Then

$u = \sum_i (\lambda_i \|x_i\|).(1/\|x_i\|.x_i) \otimes y_i$; since $\varepsilon^\infty = \ell^\infty$ we have

$$|||u|||_{(1/(1/p + 1/r), \infty, s)} \leq \| (\lambda_i \|x_i\|) \|_{\ell^{(pr)/(p+r)}} \cdot \left\| \left(\frac{x_i}{\|x_i\|} \right) \right\|_{\ell^\infty} \|(y_i)\|_{\varepsilon^s}$$

$$= \left(\sum_i |\lambda_i|^{(pr)/(p+r)} \cdot \|x_i\|^{(pr)/(p+r)} \right)^{(p+r)/pr} .1. \|(y_i)\|_{\varepsilon^s}$$

$$\leq \left(\| (|\lambda_i|^{(pr)/p+r}) \|_{\ell^{r/(p+r)}} \cdot \| (\|x_i\|^{(pr)/p+r}) \|_{\ell^{p/(p+r)}} \right)^{(p+r)/(pr)} \cdot$$

$\cdot \|(y_i)\|_{\varepsilon^s}$ by the Hölder-inequality.

$$= \|(\lambda_i)\|_{\ell^p} \ \|(x_i)\|_{\ell^r} \ \|(y_i)\|_{\varepsilon^s}.$$

Thus $|||u|||_{(1/1/p+1/r),\infty,s)} \leq |||u|||_{(\ell^p,\ell^r,\varepsilon^s)}.$

On the other hand we have

$$u = \sum_i \left(\operatorname{sign}\lambda_i \cdot |\lambda_i|^{r/(p+r)}\right)\left(|\lambda_i|^{p/(p+r)} \cdot \frac{x_i}{\|(x_j)\|_{\ell^\infty}}\right) \otimes \left(\|(x_j)\|_{\ell^\infty} \cdot y_i\right)$$

$$|||u|||_{(\ell^p,\ell^r,\varepsilon^s)} \leq \left\|\left(\operatorname{sign}\lambda_i \cdot |\lambda_i|^{r/(p+r)}\right)\right\|_{\ell^p} \cdot$$

$$\cdot \left\|\left(|\lambda_i|^{p/(p+r)} \cdot \frac{x_i}{\|(x_j)\|_{\ell^\infty}}\right)\right\|_{\ell^r} \cdot \ \|(\ \|(x_j)\|_{\ell^\infty} \cdot y_i)\|_{\varepsilon^s}$$

$$= \left(\sum_i |\lambda_i|^{(pr)/(p+r)}\right)^{1/p} \cdot \left(\sum_i |\lambda_i|^{(pr)/(p+r)} \cdot \left\|\frac{x_i}{\|(x_j)\|_{\ell^\infty}}\right\|^r\right)^{1/r}$$

$$\cdot \|(x_j)\|_{\ell^\infty} \cdot \ \|(y_i)\|_{\varepsilon^s}$$

$$\leq \left(\sum_i |\lambda_i|^{(pr)/(p+r)}\right)^{1/p} \cdot \left(\sum_i |\lambda_i|^{(pr)/(p+r)}\right)^{1/r} \cdot$$

$$\cdot \|(x_j)\|_{\ell^\infty} \cdot \|(y_i)\|_{\varepsilon^s}$$

$$= \|(\lambda_i)\|_{\ell^{1/(1/p+1/r)}} \ \|(x_i)\|_{\ell^\infty} \ \|(y_i)\|_{\varepsilon^s},$$

Thus $|||u|||_{(\ell^p,\ell^r,\varepsilon^s)} \leq |||u|||_{(1/(1/p+1/r),\infty,s)}.$

If $r = \infty$ then there is nothing to prove, the case $p = \infty$ offers no difficulties, just choose the representation

$u = \sum_i \|x_i\| \cdot \left(\lambda_i \dfrac{x_i}{\|x_i\|}\right) \otimes y_i$ and proceed as above for both

inequalities.

The third assertion follows from the first two by remarking that the form of the ε^s-norm never mattered in the proof. qed

7.9. Lemma: Let be $X, Y \in$ Ban and $1 = 1/p + 1/r + 1/s$. For all $u \in X \hat{\otimes}_{(p,r,s)} Y$ and all $\varepsilon > 0$ there exist sequences $(\lambda_i) \in \ell^p$, $(x_i) \in \varepsilon'(X)$, $(y_i) \in \varepsilon^s(Y)$ such that

$$\lim_{n\to\infty} \| u - \sum_{i=1}^{n} \lambda_i x_i \otimes y_i \|_{(p,r,s)} = 0$$

and $\|(\lambda_i)\|_{\ell^p} \|(x_i)\|_{\varepsilon^r} \|(y_i)\|_{\varepsilon^s} \leq \|u\|_{(p,r,s)} + \varepsilon$

The proof is routine calculation: take as model e.g. the proof of Proposition 3.2. in SAPHAR [20].

7.10. Lemma: $(X \hat{\otimes}_{(p,r,s)} Y)'$ is the space of all $f \in H(X,Y')$ for which there exists a $\rho > 0$ such that for all finite sequences $(\lambda_i)_{i=1}^n \subset I$, $(x_i)_{i=1}^n \subset X$, $(y_i)_{i=1}^n \subset Y$ we have

$$|\sum_i \lambda_i \langle y_i, f(x_i)\rangle| \leq \rho . \|(\lambda_i)\|_{\ell^p} \|(x_i)\|_{\varepsilon^r} \|(y_i)\|_{\varepsilon^s}.$$

Furthermore $\|f\|_{(X \hat{\otimes}_{(p,r,s)} Y)'} = \inf \rho$.

Proof: In the beginning of the proof of theorem 3.2 we gave an argument that shows that $(X \hat{\otimes}_{(p,r,s)} Y)'$ is a subspace of $H(X,Y')$, consisting of all $f \in H(X,Y')$ which induce a continuous linear functional on $X \hat{\otimes}_{(p,r,s)} Y$ by $\sum_i x_i \otimes y_i \mapsto \sum_i \langle y_i, f(x_i)\rangle$.

By that the case $q = 1$ is settled. For the general case we should bear in mind that the sup of the absolute value of a linear functional on a set coincides with the sup on the convex hull of the set (compare the proof of 7.6). qed.

7.11. Corollary: If $0 < p \leq 1$, then $\| . \|_{(p,r,s)} = \| . \|^{\wedge}$ for all r,s.

Proof: $(X \hat{\otimes}_{(p,r,s)} Y)' = H(X,Y')$, since for $f \in H(X,Y')$ we have

$$|\Sigma \lambda_i \langle y_i, f(x_i) \rangle| \leq \|f\| \, \|(\lambda_i)\|_{\ell^1} \, \|(x_i)\|_{\ell^\infty} \, \|(y_i)\|_{\ell^\infty}$$

$\leq \|f\| \, \|(\lambda_i)\|_{\ell^p} \, \|(x_i)\|_{\varepsilon^r} \, \|(y_i)\|_{\varepsilon^s}$ since $p \leq 1$, $r, s \leq \infty$

and $t \to \|(\lambda_i)\|_{\ell^t}$, $t \to \|(x_i)\|_{\varepsilon^t}$ is non increasing (compare 7.4, where we proved this).

Thus $\|f\|_{(X \otimes_{(p,r,s)} Y)'} \leq \|f\|$ and the reversed inequality holds too.
qed.

7.11. Definition: We denote the operator ideal $D \hat{\otimes}_{(p,r,s)}$ (see 4.11) by $\Pi_{(p,r,s)}$ and call it the ideal of the absolutely - (p,r,s)-summing operators, if $1 \leq 1/q = 1/p + 1/r + 1/s < \infty$, $0 < p,r,s \leq \infty$. We have then by 4.11, 3.2:

$$\Pi_{(p,r,s)}(X,Y) = D(X \hat{\otimes}_{(p,r,s)} .)(Y)$$

$$= \{f \in H(X,Y): i_y \circ f \in (X \hat{\otimes}_{(p,r,s)} Y')'\}$$

7.12. Corollary: Let be $0 < p,r,s \leq \infty$, $1 \leq 1/q = 1/p+1/r+1/s < \infty$. Then the following statements hold.

(a) Take $f \in H(X,Y)$. Then $f \in \Pi_{(p,r,s)}(X,Y)$ if and only if there is a $\rho > 0$ such that for all finite sequences $(\lambda_i) \subset I$, $(x_i) \subset X$, $(y_i') \subset Y'$ the following holds:

$$\left|\sum_i \lambda_i \langle f(x_i), y_i' \rangle\right| \leq \rho \; \|(\lambda_i)\|_{\ell^p} \cdot \|(x_i)\|_{\varepsilon^r} \cdot \|(y_i)\|_{\varepsilon^s}$$

<u>and</u> $\|f\|_{\Pi_{(p,r,s)}} = \inf \rho.$

(b) $\Pi_{(p,r,s)}$ <u>is a maximal operator ideal.</u>

(c) $(X \hat{\otimes}_{(p,r,s)} Y)' = \Pi_{(p,r,s)}(X,Y')$

(d) <u>Take</u> $f \in H(X,Y)$. <u>Then</u> $f \in \Pi_{(p,r,s)}(X,Y)$ <u>if and only if</u> $i_y \circ f \in \Pi_{(p,r,s)}(X,Y'')$; <u>furthermore</u>

$\|f\|_{\Pi_{(p,r,s)}} = \|i_y \circ f\|_{\Pi_{(p,r,s)}}$ <u>holds.</u>

(e) $\Pi_{(p,r,s)}^{\otimes} = \hat{\otimes}_{(p,r,s)}$ <u>(see 4.14).</u>

(f) $\Pi_{(p,r,s)} = H$, <u>if</u> $\rho \leq 1$ <u>for all</u> $0 < r, s \leq \infty$.

<u>Proof</u>: (a) use 7.9, (b) use 6.18, (c) use 3.3, (d) use 3.13 or 3.14, (e) use 4.14, (f) use 7.9. qed.

<u>7.13.</u> <u>Lemma</u>: <u>Let be</u> $p \geq 1$, $1/p + 1/p' = 1$, $1 \leq 1/q = 1/p + 1/r + 1/s < \infty$.

(a) <u>Take</u> $f \in H(X,Y)$. <u>Then</u> $f \in \Pi_{(p,r,s)}(X,Y)$ <u>if and only if there is a</u> $\rho > 0$ <u>such that for all finite sequences</u> $(x_i) \subset X, (y_i) \subset Y$ <u>the following holds:</u>

$$\|(\langle f(x_i), y_i' \rangle)\|_{\ell^{p'}} \leq \rho \cdot \|(x_i)\|_{\varepsilon^r} \cdot \|(y_i')\|_{\varepsilon^s}.$$

<u>Furthermore we have</u> $\|f\|_{\Pi_{(p,r,s)}} = \inf \rho.$

(b) <u>Take</u> $f \in H(X,Y)$. <u>Then</u> $f \in \Pi_{(p,p',\infty)}(X,Y)$ <u>if and only if there is a</u> $\rho > 0$ <u>such that for each finite sequence</u> $(x_i) \subset X$ <u>we have</u>

$$\|(fx_i)\|_{\ell^{p'}} \leq \rho \|(x_i)\|_{\varepsilon^{p'}}.$$

<u>Proof</u>: (a) If $f \in \Pi_{(p,r,s)}(X,Y)$, then

$$\|(\langle f(x_i),y_i'\rangle)\|_{\ell^{p'}} = \sup_{\|(\lambda_i)\|_{\ell^p}\leq 1} |\sum_i \lambda_i \langle f(x_i),y_i'\rangle|$$

$$\leq \sup_{\|(\lambda_i)\|_{\ell^p} \leq 1} \|f\|_{\Pi_{(p,r,s)}} \cdot \|(\lambda_i)\|_{\ell^p} \cdot \|(x_i)\|_{\varepsilon^r} \cdot \|(y_i')\|_{\varepsilon^s}$$

$$= \|f\|_{\Pi_{(p,r,s)}} \cdot \|(x_i)\|_{\varepsilon^r} \cdot \|(y_i')\|_{\varepsilon^s}.$$

Thus $\inf \rho \leq \|f\|_{\Pi_{(p,r,s)}}$.

If on the other hand f satisfies the condition, then for finite sequences $(\lambda_i) \subset I$, $(x_i) \subset X$, $(y_i) \subset Y'$ we have:

$$|\Sigma \lambda_i \langle f(x_i),y_i'\rangle| \leq \|(\lambda_i)\|_{\ell^p} \cdot \|(\langle f(x_i),y_i'\rangle)\|_{\ell^{p'}}$$

$$\leq (\inf \rho) \cdot \|(\lambda_i)\|_{\ell^p} \cdot \|(x_i)\|_{\varepsilon^r} \cdot \|(y_i')\|_{\varepsilon^s},$$

so $\|f\|_{\Pi_{(p,r,s)}} \leq \inf. \rho$.

(b) If $f \in \Pi_{(p,p',\infty)}(X,Y)$, then we have

$$\|(fx_i)\|_{\ell^{p'}} = \sup_{\|(\lambda_i)\|_{\ell^p}\leq 1} \left|\sum_i \lambda_i \|f(x_i)\|\right|$$

$$= \sup_{\|(\lambda_i)\|_{\ell^p}\leq 1} \left|\sum_i \lambda_i \sup_{\|y_i'\|\leq 1} |\langle f(x_i), y_i'\rangle|\right|$$

$$= \sup_{\|(\lambda_i)\|_{\ell^p}\leq 1} \sup_{\|y_i'\|\leq 1} \left|\sum_i \lambda_i \langle f(x_i), y_i'\rangle\right|$$

$$\leq \|f\|_{\Pi_{(p,p',\infty)}} \|(x_i)\|_{\varepsilon^{p'}}.$$

If f satisfies the condition, then

$$\left|\sum \lambda_i \langle f(x_i), y_i'\rangle\right| \leq \sum_i |\lambda_i| \cdot \|f(x_i)\| \cdot \|y_i'\|$$

$$\leq \|(\lambda_i)\|_{\ell^p} \cdot \|(f(x_i))\|_{\ell^{p'}} \cdot \|(y_i')\|_{\ell^\infty}$$

$$\leq (\inf \rho). \|(\lambda_i)\|_{\ell^p} \cdot \|(x_i)\|_{\varepsilon^{p'}} \cdot \|(y_i')\|_{\varepsilon^\infty}. \qquad \text{qed}$$

7.14. Remark: Let $p \geq 1$, $1/p + 1/p' = 1$, $1 \leq 1/q = 1/p + 1/r + 1/s < \infty$.

(a) Then for all X, Y the space $\Pi_{(p,r,s)}(X,Y)$ coincides isometrically with $P_{(p',r,s)}(X,Y)$, the operator ideal of "(p',r,s)-absolutely-summing" operators $X \to Y$ as defined in PIETSCH [19], 14.1.1., whose defining property is exactly the condition in 7.12(a) if $p \geq 1$. The restriction $1/q \geq 1$ corresponds exactly to the restriction $1/p' \leq 1/r + 1/s$ in [19].

(b) Lemma 7.12(b) shows that $\Pi_{(p,p',\infty)}(X,Y)$ is the operator ideal $P_{p'}(X,Y)$ of "p'-absolutely summing" operators, see e.g. PIETSCH [19], 14.3, which is well known.

<u>7.15.</u> Let be $1 \leq 1/q = 1/p + 1/r + 1/s < \infty$ and let X, Y be Banachspaces. The canonical map $s: X \hat{\otimes}_{(p,r,s)} Y \to L(X',Y)$ (see 1.12) is contractive. We consider its canonical factorisation

$$\begin{array}{ccc} X \hat{\otimes}_{(p,r,s)} Y & \xrightarrow{s} & L(X',Y) \\ \text{coims} \downarrow & & \uparrow \text{ims} \\ L^{(p,r,s)}(X',Y) & \longrightarrow & X \hat{\hat{\otimes}} Y, \end{array}$$

where $L^{(p,r,s)}(X',Y) = X \hat{\otimes}_{(p,r,s)} Y/s^{-1}(0)$ is the space of all (p,r,s)-nuclear maps which are weak $*$-norm-continuous on OX'. By duality we transform this diagram into

$$\begin{array}{ccc} (L(X',Y))' & \xrightarrow{s'} & \Pi_{(p,r,s)}(X,Y') \\ \downarrow & & \uparrow \\ I_1(X,Y') & \xrightarrow{s'} & (L^{(p,r,s)}(X',Y))' \end{array}$$

and this tells us that $I_1(X,Y')$ is contractively contained in $\Pi_{(p,r,s)}(X,Y')$. Using property 7.11(d) we conclude that $I_1(X,Y)$ is contractively contained in $\Pi_{(p,r,s)}(X,Y)$ for all X,Y. In fact I_1 is contractively contained in any maximal operator ideal Λ by exactly the same proof, using 6.17.

We have clearly $L^{(p,r,s)}(X',Y) = L^{(p,s,r)}(Y',X)$ by transposition.

7.16 We now consider the canonical factorisation of

$$s: X' \hat{\otimes}_{(p,r,s)} Y \to K(X,Y) \text{ for } 1 \leq 1/q = 1/p = 1/r = 1/s < \infty.$$

$$\begin{array}{ccc} X' \hat{\otimes}_{(p,r,s)} Y & \xrightarrow{\quad s \quad} & K(X,Y) \\ \text{coims} \downarrow & & \uparrow \text{ims} \\ N_{(p,r,s)}(X,Y) & \xrightarrow{\quad s \quad} & X' \hat{\hat{\otimes}} Y, \end{array}$$

where $N_{(p,r,s)}(X,Y) = X' \hat{\otimes}_{(p,r,s)} Y/s^{-1}(0)$

$= \operatorname{tot}(.' \hat{\otimes}_{(p,r,s)} ..)(X,Y)$ is the ideal of (p,r,s)-nuclear operators $X \to Y$.

For $q = 1$ we get the (p,r,s)-nuclear operators of PIETSCH [19], 13.1, but for $q > 1$ PIETSCH considers the quotient - q-norm of $|||\cdot|||_{(p,r,s)}$ rather than its convex hull. Clearly $N_{(p,r,s)}$ is a operator ideal for all p,r,s.

We have trivially $N_{(p,r,s)}(X,Y) = L^{(p,r,s)}(Y',X')$ by $f \to f'$ and $f' \to f'' \circ i_X$, since these maps correspond exactly to the transposition $x' \otimes y \to y \otimes x'$.

7.17. As a model for our following consideration we repeat a well known situation:

The projective tensor product $\hat{\otimes}$ is computable. $D\hat{\otimes} = H$ (cf. 4.11) is a maximal operator ideal. The tensor product $H(.',..)_e$ happens to be computable again; we call it the inductive tensor product $\hat{\hat{\otimes}}$. $D\hat{\hat{\otimes}} = I_1$, the ideal of integral operators, is a maximal operator ideal.

The tensor product $I_1(.',..)_e$ is not computable in general (in fact $[I_1(.',..)_e](X,.)$ is computable iff X has the metric approximation property). Its associated computable tensor product $L(I_1(.',..)_e) = L(I_1(.',..))$ coincides with $\hat{\otimes}$ (since $I_1(E',F) = E\hat{\otimes}F$ for $E,F \in \mathrm{Fin}$) and we arrived at the beginning again.

7.18. Now we repeat this discussion for the tensor product $\hat{\otimes}_{(p,r,s)}$, $1 \leq 1/q = 1/p + 1/r + 1/s < \infty$.

The projective (p,r,s)-tensor product is computable (cf 7.2). The ideal of (p,r,s)-absolutely summing operator $D\hat{\otimes}_{(p,r,s)} = \Pi_{(p,r,s)}$ (cf. 7.11, 7.12) is a maximal operator ideal.

We do not know whether the tensor product $\Pi_{(p,r,s)}(.',..)_e$ is computable again.

Definition: The associated computable tensor product $L(\Pi_{(p,r,s)}(.',..)_e) = L(\Pi_{(p,r,s)}(.',..))$ is called the inductive (p,r,s)-tensor product. We denote it by $\hat{\otimes}_{(p,r,s)'}$, its norm by $\| \cdot \|_{(p,r,s)'}$ since it is exactly the dual tensor norm of

$\| \cdot \|_{(p,r,s)}$ in the sense of GROTHENDIECK [8].

<u>7.19.</u> <u>Definition</u>: The operator ideal $D\hat{\otimes}_{(p,r,s)'}$ (cf.4.11) is called the ideal of (p,r,s)-integral operators. We denote it by $I_{(p,r,s)}$

By definition (4.11) we have that $I_{(p,r,s)}(X,Y)$

$$= \{f: X \to Y: i_Y \circ f \in (X \hat{\otimes}_{(p,r,s)'} Y')'\} \text{ with}$$

$$\| f \|_{I_{(p,r,s)}} = \| i_Y \circ f \|_{(X\hat{\otimes}_{(p,r,s)'} Y')'}.$$

<u>Corollary</u>: (a) $I_{(p,r,s)}$ <u>is a maximal operator ideal.</u>

(b) $(X \hat{\otimes}_{(p,r,s)'} Y)' = I_{(p,r,s)}(X,Y')$

(c) <u>Take</u> $f \in H(X,Y)$. <u>Then</u> $f \in I_{(p,r,s)}(X,Y)$ <u>if and only if</u> $i_Y \circ f \in I_{(p,r,s)}(X,Y'')$; <u>furthermore</u> $\| f \|_{I_{(p,r,s)}} = \| i_Y \circ f \|_{I_{(p,r,s)}}$ <u>in that case.</u>

(d) $I_{(p,r,s)}^{\otimes} = \hat{\otimes}_{(p,r,s)'}$

(e) $I_{(p,r,s)} = R(.' \hat{\otimes}_{(p,r,s)} ..)^{(I)}$

$$= N^{max}_{(p,r,s)}.$$

(f) <u>The tensor product</u> $I_{(p,r,s)}(.',..)_e$ <u>is not computable in general. However we have</u> $L(I_{(p,r,s)}(.',..)_e) = L(I_{(p,r,s)}(.',..))$ $= \hat{\otimes}_{(p,r,s)}$, <u>the projective</u> (p,r,s) <u>tensor product</u>.

(g) $I_{(p,r,s)}(X,Y') = I_{(p,s,r)}(Y,X')$ by transposition.

Proof: (a) use 6.16; (b) use 3.3; (c) use 3.13 or 3.14, (d) use 4.14.

(e) $I_{(p,r,s)} = D\hat{\otimes}_{(p,r,s)}'$ by definition

$= D(L\Pi_{(p,r,s)}(.',..))$ by 7.18

$= ((L\,\Pi_{(p,r,s)}(.',..'))')^{(I)}$ by 3.2, 4.11

$= ((L(\Pi_{(p,r,s)}(.',..')|Fin))')^{(I)}$

$= ((L((.'\hat{\otimes}_{(p,r,s)}..)'|Fin))')^{(I)}$ by 7.12

$= (R((.'\hat{\otimes}_{(p,r,s)}..)''|Fin))^{(I)}$ by 6.15,

or by using directly the exponential low 1.8.

$= (R(.'\hat{\otimes}_{(p,r,s)}..))^{(I)}$

$= (R(N_{(p,r,s)}|Fin))^{(I)}$

$= N_{(p,r,x)}^{max}$

(f) $L(I_{(p,r,s)}(.',..)_e) = L(I_{(p,r,s)}(.',..'')|Fin)$

$= L((.'\hat{\otimes}_{(p,r,s)}..')'|Fin)$ by (b)

$= L((\Pi_{(p,r,s)}(.,..'))'|Fin)$ by 7.18

$= L((.\hat{\otimes}_{(p,r,s)}..)''|Fin)$ by 7.12

$= \hat{\otimes}_{(p,r,s)}$

(g) trivial qed.

<u>7.20. Remark</u>: (a) If $1/q = 1$, then we remarked in 7.16, $N_{(p,r,s)}$ is exactly the space of (p,r,s)-nuclear operators in the sense of PIETSCH, [19], 13.1. Thus in the case $1/q = 1$ the ideal $I_{(p,r,s)}$ coincides exactly with the ideal $J_{(r,s)}$ of (r,s) integral operators in the sense of PIETSCH [19], 15.3 since he defines them by

$$J_{(r,s)} = N_{(p,r,s)}^{\max} = I_{(p,r,s)} \text{ by 7.19(e).}$$

(b) The adjoint of the canonical map

$X \hat{\otimes} Y \to X \hat{\otimes}_{(p,r,s)} Y$ is the embedding $\Pi_{(p,r,s)}(X,Y') \subset H(X,Y')$.

By duality thus any equality or inequality between different ideals of absolutely summing operators carries over to an equality or reversed inequality of tensor products. And since an inequality means there is a canonical map which is contractive, and an equality means: there is a canonical isomorphism, we can carry this map through all of our natural constructions 7.15, 7.16, 7.18, 7.19, thus obtaining equalities and inequalities between inductive tensor products, nuclear operators and integral operators. PIETSCH [19] collects a lot of results for ideals of absolutely summing operators in general and for special Banach spaces. We will carry them over now, combine them with our results and list all consequences for nuclear operators and integral operators too, which seem to be new. We use the convention

$1 \leq 1/q_i = 1/p_i + 1/r_i + 1/s_i < \infty$, $1/p + 1/p' = 1$ whenever $1 \leq p$. Cases $p \leq 1$ are always trivial since $\hat{\otimes}_{(p,r,s)} = \hat{\otimes}$.

"Contractively contained" for operator ideals will be denoted by $\subseteq$.

H will always denote a Hilbert space, L^t is any space $L^t(\Omega,\Sigma,\mu)$ where (Ω,Σ,μ) is a measure space, or even any abstract L^t-space for $1 \leq t \leq \infty$, c_0 is the space of null-sequences, and any result for c_0 hold for all pre-L^1-spaces too.

7.21. If $p_1 \geq p_2$, $r_1 \geq r_2$, $s_1 \geq s_2$ hold or $p_1 \geq p_2$, $r_1 \leq r_2$, $s_1 \leq s_2$, $1/q_1 \leq 1/q_2$, then we have by [19], 14.1.6 (or by direct computation):

$$\|\cdot\|_{(p_1,r_1,s_1)} \leq \|\cdot\|_{(p_2,r_2,s_2)}, \quad \Pi_{(p_1,r_1,s_1)} \subseteq \Pi_{(p_2,r_2,s_2)}$$

$$N_{(p_2,r_2,s_2)} \subseteq N_{(p_1,r_1,s_1)}, \quad L^{(p_2,r_2,s_2)} \subseteq L^{(p_1,r_1,s_1)}$$

$$\|\cdot\|_{(p_2,r_2,s_2)'} \leq \|\cdot\|_{(p_1,r_1,s_1)'}, \quad I_{(p_2,r_2,s_2)} \subseteq I_{(p_1,r_1,s_1)}$$

7.22. If $0 < r_1,r_2 < 1$, $1 \leq p_1,p_2$, $1/p_1 + 1/r_1 = 1/p_2 + 1/r_2$, then we have for all s by [19], 14.1.7:

$$\|\cdot\|_{(p_1,r_1,s)} = \|\cdot\|_{(p_2,r_2,s)}, \quad \Pi_{(p_1,r_1s)} = \Pi_{(p_2,r_2,s)}$$

$$N_{(p_1,r_1,s)} = N_{(p_2,r_2,s)} \quad L^{(p_1,r_1,s)} = L^{(p_2,r_2,s)}$$

$$\|\cdot\|_{(p_1,r_1,s)'} = \|\cdot\|_{(p_2,r_2,s)'}, \quad I_{(p_1,r_1,s)} = I_{(p_2,r_2s)},$$

and the same statements hold if we interchange r_1,r_2 and s.

7.23. $\|\cdot\|_{(\infty,1,1)} = \|\cdot\|^{\wedge}$, $\Pi_{(\infty,1,1)} = H$

$N_{(\infty,1,1)} = N_1$, $L_{(\infty,1,1)} = L^1$

$\|\cdot\|_{(\infty,1,1)'} = \|\cdot\|^{\hat{\wedge}}$, $I_{(\infty,1,1)} = I_1$ by [19], 14.1.9.

7.24. If $1/q_1 + 1/q_2 = 1$ and $r_1 \leq r_2$, $s_1 \leq s_2$, then

$$\|\cdot\|_{(p_1,r_1,s_1)} \leq \|\cdot\|_{(p_2,r_2,s_2)}, \quad \Pi_{(p_1,r_1,s_1)} \subseteq \Pi_{(p_2,r_2,s_2)}$$

$$N_{(p_2,r_2,s_2)} \subseteq N_{(p_1,r_1,s_1)}, \quad L^{(p_2,r_2,s_2)} \subseteq L^{(p_1,r_1,s_1)}$$

$$\|\cdot\|_{(p_2,r_2,s_2)} \leq \|\cdot\|_{(p_1,r_1,s_1)}, \quad I_{(p_2,r_2,s_2)} \subseteq I_{(p_1,r_1,s_1)})$$

by [19], 14.4.2. I wonder whether $1/q_1 \geq 1/q_2$, $r_1 \leq r_2$, $s_1 \leq s_2$ suffices.

7.25. If $1/q = 1$, then by [19], 14.4.3 we have

$$\|\cdot\|_{(p,r,s)} \leq \|\cdot\|_{(\infty,2,2)}, \quad \Pi_{(p,r,s)} \subseteq \Pi_{(\infty,2,2)}$$

$$N_{(\infty 2,2)} \subseteq N_{(p,r,s)}, \quad L^{(\infty,2,2)} \subseteq N^{(p,r,s)}$$

$$\|\cdot\|_{(\infty,2,2)'} \leq \|\cdot\|_{(p,r,s)'}, \quad I_{(\infty,2,2)} \subseteq I_{(p,r,s)},$$

and we have equality everywhere whenever $2 \leq r, s, < \infty$.

7.26. Each of the following conditions implies, that

$$\| \cdot \|_{(p,r,s)} = \| \cdot \|^{\wedge}, \quad \Pi_{(p,r,s)} = H$$

$$N_{(p,r,s)} = N_1, \quad L^{(p,r,s)} = L^1,$$

$$\| \cdot \|_{(p,r,s)'} = \| \cdot \|^{\hat{\wedge}} ; \; I_{(p,r,s)} = I_1 ;$$

(a) $p \leq 1$; r,s arbitrary, by 7.10

(b) $0 < r < 1$, $1/p + 1/r > 2$, s arbitrary.

Then the equation $1 + 1/x = 1/p + 1/r$ has a solution $0 < x < 1$ and we can apply 7.22 and (a).

(c) $0 < s < 1$, $1/p + 1/s > 2$, r arbitrary.

(d) p arbitrary, $0 < r$, $s \leq 1$, by 7.21 and 7.23.

7.27. If $1/q=1$, r, $s \neq \infty$, then we have by [19], 14.5.1

$$H \hat{\hat{\otimes}}_{(p,r,s)} H = N_{(p,r,s)}(H,H) = L^{(p,r,s)}(H,H) = H \hat{\hat{\otimes}} H = K(H,H),$$

$$\Pi_{(p,r,s)}(H,H) = H \hat{\otimes}_{(p,r,s)'} H = H \hat{\otimes} H = N_1(H,H) = I_1(H,H)$$

$$I_{(p,r,s)}(H,H) = H(H,H) = \mathfrak{B}(H)$$

If $1 < p < \infty$, then we have

$$H \hat{\otimes}_{(p,p',\infty)} H = \Pi_{(p,p',\infty)}(H,H) = N_{(p,p',\infty)}(H,H)$$

$$= L^{(p,p',\infty)}(H,H) = H \hat{\otimes}_{(p,p',\infty)'} H = I_{(p,p',\infty)}(H,H)$$

$= \mathfrak{S}_2$, the space of Hilbert-Schmidt operators. The same equations hold, if we exchange (p,∞,p') for (p,p',∞).

7.28. $H \hat{\otimes}_{(\infty,2,2)} H = N_{(\infty,2,2)}(H,H) = L^{(\infty,2,2)}(H,H) = H \hat{\hat{\otimes}} H = K(H,H)$,

$\Pi_{(\infty,2,2)}(H,H) = H \hat{\otimes}_{(\infty,2,2)'} H = H \hat{\otimes} H = N_1(H,H)$.

$I_{(\infty,2,2)}(H,H) = H(H,H)$.

If $1 \leq p < \infty$, then we have:

$H \hat{\otimes}_{(p,2,2)} H = N_{(p,2,2)}(H,H) = L^{(p,2,2)}(H,H)$

$= I_{(p,2,2)}(H,H) = \mathfrak{S}_p$,

$\Pi_{(p,2,2)}(H,H) = H \hat{\otimes}_{(p,2,2)'} H = \mathfrak{S}_{p'}$, where $\mathfrak{S}_p$ is the ideal of all operators in $\mathfrak{B}(H)$ whose sequences of s-numbers lie in ℓ^p.

The same equations hold if we set $(p,\infty,2)$ or $(p,2,\infty)$ for $(p,2,2)$; See [19] 14.5.2.

7.29 If $p \geq 1$ and $1 \leq r \leq 2$, then the case $1/t = 1/p + 1/r - 1/2 < 1$ implies, that the following holds, by [19], 14.5.3.

$H \hat{\otimes}_{(p,r,\infty)} H = N_{(p,r,\infty)}(H,H) = L^{(p,r,\infty)}(H,H)$

$= I_{(p,r,\infty)}(H,H) = \mathfrak{S}_t$.

$\Pi_{(p,r,\infty)}(H,H) = H \hat{\otimes}_{(p,r,\infty)'} H = \mathfrak{S}_{t'}$.

The other case $1/p + 1/r - 1/2 \geq 1$ implies that

$H \hat{\otimes}_{(p,r,\infty)} H = N_{(p,r,\infty)}(H,H) = L^{(p,r,\infty)}(H,H)$

$= I_{(p,r,\infty)}(H,H) = H \hat{\otimes} H = N_1(H,H)$,

$\Pi_{(p,r,\infty)}(H,H) = H(H,H)$, $H \hat{\otimes}_{(p,r,\infty)'} H = H \hat{\hat{\otimes}} H$.

7.30 If X is any Banach space, $1 < p \leq 2$, $2 \leq t < \infty$, then we have by [19], 16.1.3:

$$X \hat{\otimes}_{(p,p',\infty)} L^t = L^t \hat{\otimes}_{(p,\infty,p')} X = N_{(p,\infty,p')}(L^{t'},X)$$

$$= L^{(p,p',\infty)}(X',L^t) = X \hat{\otimes}_{(2,2,\infty)} L^t,$$

$$\Pi_{(p,p',\infty)}(X,L^{t'}) = \Pi_{(p,\infty,p')}(L^t,X') = \Pi_{(2,2,\infty)}(X,L^{t'}),$$

If X has the metric approximation property, then we have furthermore,

$$X \hat{\otimes}_{(p,p',\infty)'} L^{t'} = L^{t'} \hat{\otimes}_{(p,\infty,p')} X = X \hat{\otimes}_{(2,2,\infty)} L^{t'}$$

$$I_{(p,p',\infty)}(X,L^{t'}) = I_{(2,2,\infty)}(X,L^{t'}),$$

since then we have $X \hat{\otimes}_{(p,p',\infty)'} L^{t'} = \Pi_{(p,p',\infty)}(.',..)_e(X,L^{t'})$.

7.31 The following spaces are independent of p in the intervals which we will list after then, by [19],16.1.4:

$$L^u \hat{\otimes}_{(p,p',\infty)} L^v = L^v \hat{\otimes}_{(p,\infty.p')} L^u$$

$$= N_{(p,p',\infty)}(L^{u'},L^v) = L^{(p,p',\infty)}(L^{u'},L^v),$$

$$\Pi_{(p,p',\infty)}(L^u,L^{v'}) = \Pi_{(p,\infty,p')}(L^v,L^{u'})$$

$$L^{u'} \hat{\otimes}_{(p,p',\infty)'} L^{v'} = L^{v'} \otimes_{(p,\infty,p')'} L^{u'}$$

$$I_{(p,p',\infty)}(L^{u'},L^v) = I_{(p,\infty,p')}(L^{v'},L^u).$$

The intervals are the following ones:

$1 < p < \infty$ for $1 \leq u \leq 2$, $2 \leq v < \infty$;

$u < p \leq \infty$, $1 < p \leq 2$ for $2 \leq u < \infty, 2 \leq v < \infty$

$2 \leq p \leq \infty$ for $1 \leq u \leq 2$, $1 < v \leq 2$

$u < p \leq \infty$ for $2 \leq u < \infty$, $1 < v \leq 2$.

In some cases one can replace $1 <\ldots$ and $\ldots < \infty$ by $\leq$ with a little caution.

7.32. By [19], 16.1.5 we have:

$L^1 \hat{\otimes}_{(\infty,1,\infty)} L^2 = L^2 \hat{\otimes}_{(\infty,\infty,1)} L^1 = N_{(\infty,\infty,1)}(L^2,L^1)$

$= L^{(\infty,1,\infty)}(L^\infty,L^2) = L^1 \hat{\otimes} L^2,$

$\Pi_{(\infty,1,\infty)}(L^1,L^2) = \Pi_{(\infty,\infty,1)}(L^2,L^\infty) = H(L^2,L^\infty).$

$L^\infty \otimes_{(\infty,1,\infty)'} L^2 = L^\infty \hat{\hat{\otimes}} L^2$

$I_{(\infty,1,\infty)}(L^\infty,L^2) = I_1(L^\infty,L^2).$

7.33. If either $2 = p \leq t < \infty$, $r = 2$,

or $1 \leq p < t < 2$, $r = p'$ ([19], 16.1.7)

or $1 < p = t < r' \leq 2$ ([19], 14.1.8), then we have:

$L^\infty \hat{\otimes}_{(p,r,\infty)} L^t = L^t \hat{\otimes}_{(p,\infty,r)} L^\infty = N_{(p,r,\infty)}(L^1,L^t)$

$= L^{(p,\infty,r)}(L^{t'},L^\infty) = L^\infty \hat{\otimes} L^t,$

$\Pi_{(p,r,\infty)}(L^\infty,L^{t'}) = \Pi_{(p,\infty,r)}(L^t,(L^\infty)') = H(L^\infty,L^{t'}),$

so $\Pi_{(p,\infty,r)}(L^t,L^1) = H(L^t,L^1)$ by 7.12(d) and

$L^{t'} \hat{\otimes}_{(p,\infty,r)'} L^1 = L^{t'} \hat{\hat{\otimes}} L^1$, $I_{(p,\infty,r)}(L^{t'},L^\infty) = I_1(L^{t'},L^\infty).$

7.34: Out of 16.1.9 we can deduce:

$\Pi_{(\infty,2,2)}(L_\infty,L_1) = H(L^\infty,L^1)$, thus

$L^\infty \hat{\otimes}_{(\infty,2,2)'} L^\infty = L^\infty \hat{\hat{\otimes}} L^\infty$, $I_{(\infty,2,2)}(L^\infty,(L^\infty)') = I_1(L^\infty,(L^\infty)')$

and by 7.19(c): $I_{(\infty,2,2)}(L^\infty,L^1) = I_1(L^\infty,L^1)$;

$L^\infty \hat{\otimes}_{(\infty,2,2)} c_0 = L^\infty \hat{\otimes} c_0.$

$\Pi_{(2,2,2)}(L^1,L^1) = H(L^1,L^1)$, thus

$L^\infty \hat{\otimes}_{(2,2,2)'} L^1 = L^\infty \hat{\hat{\otimes}} L^1$, $I_{(2,2,2)}(L^\infty,L^\infty) = I_1(L^\infty,L^\infty)$

$L^1 \hat{\otimes}_{(2,2,2)} c_o = L^1 \hat{\otimes} c_o$.

$\Pi_{(2,2,2)}(L^\infty,L^\infty) = H(L^\infty,L\ )$, thus

$L^\infty \hat{\otimes}_{(2,2,2)} L^1 = L^\infty \hat{\otimes} L^1$, $(L^\infty)' \hat{\otimes}_{(2,2,2)'} L^\infty = (L^\infty)' \hat{\hat{\otimes}} L^\infty$,

$I_{(2,2,2)}((L^\infty)', (L^\infty)') = I_1((L^\infty)', (L^\infty)')$, by 7.19(c)

$I_{(2,2,2)}((L^\infty)', L^1) = I_1((L^\infty)',L^1)$.

$\Pi_{(2,1,\infty)}(L^t,L^t) = H(L^t,L^t)$, $1 \leq t \leq 2$, thus

$L^t \hat{\otimes}_{(2,1,\infty)} L^{t'} = L^t \hat{\otimes} L^{t'}$, $1 < t \leq 2$

$L^{t'} \hat{\otimes}_{(2,1,\infty)'} L^t = L^{t'} \hat{\hat{\otimes}} L^t$, $1 \leq t \leq 2$

$I_{(2,1,\infty)}(L^{t'},L^{t'}) = I_1(L^{t'},L^{t'})$, $1 \leq t \leq 2$.

$\Pi_{(t,1,\infty)}(L^t,L^t) = H(L^t,L^t)$, $2 \leq t \leq \infty$, thus

$L^t \hat{\otimes}_{(t,1,\infty)} L^{t'} = L^t \hat{\otimes} L^{t'}$, $2 \leq t \leq \infty$,

$L^{t'} \hat{\otimes}_{(t,1,\infty)'} L^t = L^{t'} \hat{\hat{\otimes}} L^t$, $2 \leq t < \infty$,

$I_{(t,1,\infty)}(L^{t'},L^{t'}) = I_1(L^{t'},L^{t'})$, $2 \leq t < \infty$

We could write M for $(L^\infty)'$, meaning an abstract M space.

REFERENCES

1 I. AMEMIGA, K. SHIGA : On tensor products of Banach spaces, Kotai Math. Sem. Rep. (1957), 161-178.

2 S. CHEVET: Sur certaines produits tensoriels topologiques d'espaces de Banach, Z. Wahrscheinlichkeitstheorie, verw. Geb. 11 (1969), 120-138.

3 J. CIGLER : Funktoren auf Kategorien von Banachraumen, Monatshefte Math.

4 J. CIGLER : Duality for functors on Banach spaces, Preprint 1973

5 J. CIGLER : Funktoren auf Kategorien von Banachraumen, Lecture Notes, University of Vienna, 1974.

6 Y. GORDON, D.R. LEWIS, J.R. RETHERFORD : Banach ideals of operators with applications, J. Functional Analysis 14, (1973), 85-92)

7 A. GROTHENDIECK : Produits tensoriels topologiques et espaces nucleaires, Mem AMS 16 (1955)

8 A. GROTHENDIECK : Résumé de la théorie metrique des produits tensoriels topologiques, Bol. Soc. Matem. Sao Paulo 7 (1952)

9 C. HERZ, J. WICK PELLETIER, Dual functors and integral operators in the category of Banach spaces, preprint 1974.

10 V.L. LEVIN : Tensor products and functors in categories of Banach spaces defined by KB-lineals,
Trudy Moscov. Mat. Obsc. 20 (1969) (Russian)
Transl. Moscov. Math. Soc. (AMS) <u>20</u> (1969), 41-77.

11 F.E.J. LINTON : Autonomous categories and duality of functors, J. Algebra 2, 315-349 (1965).

12 S. MACLANE : Categories for the working mathematician, Graduate texts in mathematics 5, Springer 1975.

13 P. MICHOR : Zum Tensorprodukt von Funktoren auf Kategorien von Banachraumen, Monatshefte Math , 78(1974), 117-130

14 P. MICHOR : Funktoren zwischen Kategorien von Banach- und Waelbroeck-Räumen, Situngsberichte Österr. Akad. Wiss. II, 182 (1973), 43-65.

15 P. MICHOR : Duality for contravariant functors on Banach spaces, preprint 1974.

16 B.S. MITJAGIN, A.S. SHVARTS : Functors on categories of Banach spaces, Russ. Math. Surveys 19 (1964), 65-127.

17 J.W. NEGREPONTIS : Duality of functors on categories of Banach spaces, J. pure appl. Algebra 3 (1973) 119-131.

18 A. PIETSCH : Adjungierte normierte Operatorenideale, Math. Nachrichten 48 (1971) 189 - 212.

19 A. PIETSCH : Theorie der Operatorenideale, Jena 1972.

20 P. SAPHAR : Produits tensoriels d'espaces de Banach et classes d'applications linéaires, Studia Math. 38 (1970) 71-100.

21 R. SCHATTEN : A theory of cross spaces, Ann. of Math. studies 26, Princeton 1950.

22 Z. SEMADENI, A. WIWEGER : A theorem of Eilenberg-Watts type for tensor products of Banach spaces, Studia Math. 38 (1970), 235-242.

23 P. ENFLO : A counter example to the approximation problem, Acta Math. 130 (1973) 309 - 317.

Peter W. Michor,

Mathematisches Institut der Universitat,

A-1090 Wien, Strudlhofgasse 4, Austria.

Vol. 489: J. Bair and R. Fourneau, Etude Géométrique des Espaces Vectoriels. Une Introduction. VII, 185 pages. 1975.

Vol. 490: The Geometry of Metric and Linear Spaces. Proceedings 1974. Edited by L. M. Kelly. X, 244 pages. 1975.

Vol. 491: K. A. Broughan, Invariants for Real-Generated Uniform Topological and Algebraic Categories. X, 197 pages. 1975.

Vol. 492: Infinitary Logic: In Memoriam Carol Karp. Edited by D. W. Kueker. VI, 206 pages. 1975.

Vol. 493: F. W. Kamber and P. Tondeur, Foliated Bundles and Characteristic Classes. XIII, 208 pages. 1975.

Vol. 494: A Cornea and G. Licea. Order and Potential Resolvent Families of Kernels. IV, 154 pages. 1975.

Vol. 495: A. Kerber, Representations of Permutation Groups II. V, 175 pages. 1975.

Vol. 496: L. H. Hodgkin and V. P. Snaith, Topics in K-Theory. Two Independent Contributions. III, 294 pages. 1975.

Vol. 497: Analyse Harmonique sur les Groupes de Lie. Proceedings 1973–75. Edité par P. Eymard et al. VI, 710 pages. 1975.

Vol. 498: Model Theory and Algebra. A Memorial Tribute to Abraham Robinson. Edited by D. H. Saracino and V. B. Weispfenning. X, 463 pages. 1975.

Vol. 499: Logic Conference, Kiel 1974. Proceedings. Edited by G. H. Müller, A. Oberschelp, and K. Potthoff. V, 651 pages 1975.

Vol. 500: Proof Theory Symposion, Kiel 1974. Proceedings. Edited by J. Diller and G. H. Müller. VIII, 383 pages. 1975.

Vol. 501: Spline Functions, Karlsruhe 1975. Proceedings. Edited by K. Böhmer, G. Meinardus, and W. Schempp. VI, 421 pages. 1976.

Vol. 502: János Galambos, Representations of Real Numbers by Infinite Series. VI, 146 pages. 1976.

Vol. 503: Applications of Methods of Functional Analysis to Problems in Mechanics. Proceedings 1975. Edited by P. Germain and B. Nayroles. XIX, 531 pages. 1976.

Vol. 504: S. Lang and H. F. Trotter, Frobenius Distributions in GL_2-Extensions. III, 274 pages. 1976.

Vol. 505: Advances in Complex Function Theory. Proceedings 1973/74. Edited by W. E. Kirwan and L. Zalcman. VIII, 203 pages. 1976.

Vol. 506: Numerical Analysis, Dundee 1975. Proceedings. Edited by G. A. Watson. X, 201 pages. 1976.

Vol. 507: M. C. Reed, Abstract Non-Linear Wave Equations. VI, 128 pages. 1976.

Vol. 508: E. Seneta, Regularly Varying Functions. V, 112 pages. 1976.

Vol. 509: D. E. Blair, Contact Manifolds in Riemannian Geometry. VI, 146 pages. 1976.

Vol. 510: V. Poènaru, Singularités C^∞ en Présence de Symétrie. V, 174 pages. 1976.

Vol. 511: Séminaire de Probabilités X. Proceedings 1974/75. Edité par P. A. Meyer. VI, 593 pages. 1976.

Vol. 512: Spaces of Analytic Functions, Kristiansand, Norway 1975. Proceedings. Edited by O. B. Bekken, B. K. Øksendal, and A. Stray. VIII, 204 pages. 1976.

Vol. 513: R. B. Warfield, Jr. Nilpotent Groups. VIII, 115 pages. 1976.

Vol. 514: Séminaire Bourbaki vol. 1974/75. Exposés 453 – 470. IV, 276 pages. 1976.

Vol. 515: Bäcklund Transformations. Nashville, Tennessee 1974. Proceedings. Edited by R. M. Miura. VIII, 295 pages. 1976.

Vol. 516: M. L. Silverstein, Boundary Theory for Symmetric Markov Processes. XVI, 314 pages. 1976.

Vol. 517: S. Glasner, Proximal Flows. VIII, 153 pages. 1976.

Vol. 518: Séminaire de Théorie du Potentiel, Proceedings Paris 1972–1974. Edité par F. Hirsch et G. Mokobodzki. VI, 275 pages. 1976.

Vol. 519: J. Schmets, Espaces de Fonctions Continues. XII, 150 pages. 1976.

Vol. 520: R. H. Farrell, Techniques of Multivariate Calculation. X, 337 pages. 1976.

Vol. 521: G. Cherlin, Model Theoretic Algebra – Selected Topics. IV, 234 pages. 1976.

Vol. 522: C. O. Bloom and N. D. Kazarinoff, Short Wave Radiation Problems in Inhomogeneous Media: Asymptotic Solutions. V. 104 pages. 1976.

Vol. 523: S. A. Albeverio and R. J. Høegh-Krohn, Mathematical Theory of Feynman Path Integrals. IV, 139 pages. 1976.

Vol. 524: Séminaire Pierre Lelong (Analyse) Année 1974/75. Edité par P. Lelong. V, 222 pages. 1976.

Vol. 525: Structural Stability, the Theory of Catastrophes, and Applications in the Sciences. Proceedings 1975. Edited by P. Hilton. VI, 408 pages. 1976.

Vol. 526: Probability in Banach Spaces. Proceedings 1975. Edited by A. Beck. VI, 290 pages. 1976.

Vol. 527: M. Denker, Ch. Grillenberger, and K. Sigmund, Ergodic Theory on Compact Spaces. IV, 360 pages. 1976.

Vol. 528: J. E. Humphreys, Ordinary and Modular Representations of Chevalley Groups. III, 127 pages. 1976.

Vol. 529: J. Grandell, Doubly Stochastic Poisson Processes. X, 234 pages. 1976.

Vol. 530: S. S. Gelbart, Weil's Representation and the Spectrum of the Metaplectic Group. VII, 140 pages. 1976.

Vol. 531: Y.-C. Wong, The Topology of Uniform Convergence on Order-Bounded Sets. VI, 163 pages. 1976.

Vol. 532: Théorie Ergodique. Proceedings 1973/1974. Edité par J.-P. Conze and M. S. Keane. VIII, 227 pages. 1976.

Vol. 533: F. R. Cohen, T. J. Lada, and J. P. May, The Homology of Iterated Loop Spaces. IX, 490 pages. 1976.

Vol. 534: C. Preston, Random Fields. V, 200 pages. 1976.

Vol. 535: Singularités d'Applications Differentiables. Plans-sur-Bex. 1975. Edité par O. Burlet et F. Ronga. V, 253 pages. 1976.

Vol. 536: W. M. Schmidt, Equations over Finite Fields. An Elementary Approach. IX, 267 pages. 1976.

Vol. 537: Set Theory and Hierarchy Theory. Bierutowice, Poland 1975. A Memorial Tribute to Andrzej Mostowski. Edited by W. Marek, M. Srebrny and A. Zarach. XIII, 345 pages. 1976.

Vol. 538: G. Fischer, Complex Analytic Geometry. VII, 201 pages. 1976.

Vol. 539: A. Badrikian, J. F. C. Kingman et J. Kuelbs, Ecole d'Eté de Probabilités de Saint Flour V-1975. Edité par P.-L. Hennequin. IX, 314 pages. 1976.

Vol. 540: Categorical Topology, Proceedings 1975. Edited by E. Binz and H. Herrlich. XV, 719 pages. 1976.

Vol. 541: Measure Theory, Oberwolfach 1975. Proceedings. Edited by A. Bellow and D. Kölzow. XIV, 430 pages. 1976.

Vol. 542: D. A. Edwards and H. M. Hastings, Čech and Steenrod Homotopy Theories with Applications to Geometric Topology. VII, 296 pages. 1976.

Vol. 543: Nonlinear Operators and the Calculus of Variations, Bruxelles 1975. Edited by J. P. Gossez, E. J. Lami Dozo, J. Mawhin, and L. Waelbroeck, VII, 237 pages. 1976.

Vol. 544: Robert P. Langlands, On the Functional Equations Satisfied by Eisenstein Series. VII, 337 pages. 1976.

Vol. 545: Noncommutative Ring Theory. Kent State 1975. Edited by J. H. Cozzens and F. L. Sandomierski. V, 212 pages. 1976.

Vol. 546: K. Mahler, Lectures on Transcendental Numbers. Edited and Completed by B. Diviš and W. J. Le Veque. XXI, 254 pages. 1976.

Vol. 547: A. Mukherjea and N. A. Tserpes, Measures on Topological Semigroups: Convolution Products and Random Walks. V, 197 pages. 1976.

Vol. 548: D. A. Hejhal, The Selberg Trace Formula for PSL (2, ℝ). Volume I. VI, 516 pages. 1976.

Vol. 549: Brauer Groups, Evanston 1975. Proceedings. Edited by D. Zelinsky. V, 187 pages. 1976.

Vol. 550: Proceedings of the Third Japan – USSR Symposium on Probability Theory. Edited by G. Maruyama and J. V. Prokhorov. VI, 722 pages. 1976.

Vol. 551: Algebraic K-Theory, Evanston 1976. Proceedings. Edited by M. R. Stein. XI, 409 pages. 1976.

Vol. 552: C. G. Gibson, K. Wirthmüller, A. A. du Plessis and E. J. N. Looijenga. Topological Stability of Smooth Mappings. V, 155 pages. 1976.

Vol. 553: M. Petrich, Categories of Algebraic Systems. Vector and Projective Spaces, Semigroups, Rings and Lattices. VIII, 217 pages. 1976.

Vol. 554: J. D. H. Smith, Mal'cev Varieties. VIII, 158 pages. 1976.

Vol. 555: M. Ishida, The Genus Fields of Algebraic Number Fields. VII, 116 pages. 1976.

Vol. 556: Approximation Theory. Bonn 1976. Proceedings. Edited by R. Schaback and K. Scherer. VII, 466 pages. 1976.

Vol. 557: W. Iberkleid and T. Petrie, Smooth S^1 Manifolds. III, 163 pages. 1976.

Vol. 558: B. Weisfeiler, On Construction and Identification of Graphs. XIV, 237 pages. 1976.

Vol. 559: J.-P. Caubet, Le Mouvement Brownien Relativiste. IX, 212 pages. 1976.

Vol. 560: Combinatorial Mathematics, IV, Proceedings 1975. Edited by L. R. A. Casse and W. D. Wallis. VII, 249 pages. 1976.

Vol. 561: Function Theoretic Methods for Partial Differential Equations. Darmstadt 1976. Proceedings. Edited by V. E. Meister, N. Weck and W. L. Wendland. XVIII, 520 pages. 1976.

Vol. 562: R. W. Goodman, Nilpotent Lie Groups: Structure and Applications to Analysis. X, 210 pages. 1976.

Vol. 563: Séminaire de Théorie du Potentiel. Paris, No. 2. Proceedings 1975–1976. Edited by F. Hirsch and G. Mokobodzki. VI, 292 pages. 1976.

Vol. 564: Ordinary and Partial Differential Equations, Dundee 1976. Proceedings. Edited by W. N. Everitt and B. D. Sleeman. XVIII, 551 pages. 1976.

Vol. 565: Turbulence and Navier Stokes Equations. Proceedings 1975. Edited by R. Temam. IX, 194 pages. 1976.

Vol. 566: Empirical Distributions and Processes. Oberwolfach 1976. Proceedings. Edited by P. Gaenssler and P. Révész. VII, 146 pages. 1976.

Vol. 567: Séminaire Bourbaki vol. 1975/76. Exposés 471–488. IV, 303 pages. 1977.

Vol. 568: R. E. Gaines and J. L. Mawhin, Coincidence Degree, and Nonlinear Differential Equations. V, 262 pages. 1977.

Vol. 569: Cohomologie Etale SGA 4½. Séminaire de Géométrie Algébrique du Bois-Marie. Edité par P. Deligne. V, 312 pages. 1977.

Vol. 570: Differential Geometrical Methods in Mathematical Physics, Bonn 1975. Proceedings. Edited by K. Bleuler and A. Reetz. VIII, 576 pages. 1977.

Vol. 571: Constructive Theory of Functions of Several Variables, Oberwolfach 1976. Proceedings. Edited by W. Schempp and K. Zeller. VI, 290 pages. 1977

Vol. 572: Sparse Matrix Techniques, Copenhagen 1976. Edited by V. A. Barker. V, 184 pages. 1977.

Vol. 573: Group Theory, Canberra 1975. Proceedings. Edited by R. A. Bryce, J. Cossey and M. F. Newman. VII, 146 pages. 1977.

Vol. 574: J. Moldestad, Computations in Higher Types. IV, 203 pages. 1977.

Vol. 575: K-Theory and Operator Algebras, Athens, Georgia 1975. Edited by B. B. Morrel and I. M. Singer. VI, 191 pages. 1977.

Vol. 576: V. S. Varadarajan, Harmonic Analysis on Real Reductive Groups. VI, 521 pages. 1977.

Vol. 577: J. P. May, E_∞ Ring Spaces and E_∞ Ring Spectra. IV, 268 pages. 1977.

Vol. 578: Séminaire Pierre Lelong (Analyse) Année 1975/76. Edité par P. Lelong. VI, 327 pages. 1977.

Vol. 579: Combinatoire et Représentation du Groupe Symétrique, Strasbourg 1976. Proceedings 1976. Edité par D. Foata. IV, 339 pages. 1977.

Vol. 580: C. Castaing and M. Valadier, Convex Analysis and Measurable Multifunctions. VIII, 278 pages. 1977.

Vol. 581: Séminaire de Probabilités XI, Université de Strasbourg. Proceedings 1975/1976. Edité par C. Dellacherie, P. A. Meyer et M. Weil. VI, 574 pages. 1977.

Vol. 582: J. M. G. Fell, Induced Representations and Banach *-Algebraic Bundles. IV, 349 pages. 1977.

Vol. 583: W. Hirsch, C. C. Pugh and M. Shub, Invariant Manifolds. IV, 149 pages. 1977.

Vol. 584: C. Brezinski, Accélération de la Convergence en Analyse Numérique. IV, 313 pages. 1977.

Vol. 585: T. A. Springer, Invariant Theory. VI, 112 pages. 1977.

Vol. 586: Séminaire d'Algèbre Paul Dubreil, Paris 1975–1976 (29ème Année). Edited by M. P. Malliavin. VI, 188 pages. 1977.

Vol. 587: Non-Commutative Harmonic Analysis. Proceedings 1976. Edited by J. Carmona and M. Vergne. IV, 240 pages. 1977.

Vol. 588: P. Molino, Théorie des G-Structures: Le Problème d'Equivalence. VI, 163 pages. 1977.

Vol. 589: Cohomologie l-adique et Fonctions L. Séminaire de Géométrie Algébrique du Bois-Marie 1965–66, SGA 5. Edité par L. Illusie. XII, 484 pages. 1977.

Vol. 590: H. Matsumoto, Analyse Harmonique dans les Systèmes de Tits Bornologiques de Type Affine. IV, 219 pages. 1977.

Vol. 591: G. A. Anderson, Surgery with Coefficients. VIII, 157 pages. 1977.

Vol. 592: D. Voigt, Induzierte Darstellungen in der Theorie der endlichen, algebraischen Gruppen. V, 413 Seiten. 1977.

Vol. 593: K. Barbey and H. König, Abstract Analytic Function Theory and Hardy Algebras. VIII, 260 pages. 1977.

Vol. 594: Singular Perturbations and Boundary Layer Theory, Lyon 1976. Edited by C. M. Brauner, B. Gay, and J. Mathieu. VIII, 539 pages. 1977.

Vol. 595: W. Hazod, Stetige Faltungshalbgruppen von Wahrscheinlichkeitsmaßen und erzeugende Distributionen. XIII, 157 Seiten. 1977.

Vol. 596: K. Deimling, Ordinary Differential Equations in Banach Spaces. VI, 137 pages. 1977.

Vol. 597: Geometry and Topology, Rio de Janeiro, July 1976. Proceedings. Edited by J. Palis and M. do Carmo. VI, 866 pages. 1977.

Vol. 598: J. Hoffmann-Jørgensen, T. M. Liggett et J. Neveu, Ecole d'Eté de Probabilités de Saint-Flour VI – 1976. Edité par P.-L. Hennequin. XII, 447 pages. 1977.

Vol. 599: Complex Analysis, Kentucky 1976. Proceedings. Edited by J. D. Buckholtz and T. J. Suffridge. X, 159 pages. 1977.

Vol. 600: W. Stoll, Value Distribution on Parabolic Spaces. VIII, 216 pages. 1977.

Vol. 601: Modular Functions of one Variable V, Bonn 1976. Proceedings. Edited by J.-P. Serre and D. B. Zagier. VI, 294 pages. 1977.

Vol. 602: J. P. Brezin, Harmonic Analysis on Compact Solvmanifolds. VIII, 179 pages. 1977.

Vol. 603: B. Moishezon, Complex Surfaces and Connected Sums of Complex Projective Planes. IV, 234 pages. 1977.

Vol. 604: Banach Spaces of Analytic Functions, Kent, Ohio 1976. Proceedings. Edited by J. Baker, C. Cleaver and Joseph Diestel. VI, 141 pages. 1977.

Vol. 605: Sario et al., Classification Theory of Riemannian Manifolds. XX, 498 pages. 1977.

Vol. 606: Mathematical Aspects of Finite Element Methods. Proceedings 1975. Edited by I. Galligani and E. Magenes. VI, 362 pages. 1977.

Vol. 607: M. Métivier, Reelle und Vektorwertige Quasimartingale und die Theorie der Stochastischen Integration. X, 310 Seiten. 1977.

Vol. 608: Bigard et al., Groupes et Anneaux Réticulés. XIV, 334 pages. 1977.